KV-372-306

Our Dynamic World

Economic Activities (Elective)

Barry Brunt

> **My-etest**
>
> Packed full of extra questions, **my-etest** lets you revise –
> at your own pace – when you want – where you want.
> Test yourself on our FREE website **www.my-etest.com**
> and check out how well you score!
>
> **Teachers!**
> Print an etest and give it for homework or a class test.

GILL & MACMILLAN

18651331

Gill & Macmillan Ltd
Hume Avenue
Park West
Dublin 12
with associated companies throughout the world
www.gillmacmillan.ie

© Barry Brunt 2004
0 7171 3519 5
Design, colour illustrations and print origination in Ireland by Design Image, Dublin
Colour reproduction by Ultragraphics, Dublin

The paper used in this book is made from the wood pulp of managed forests. For every tree felled, at least one tree is planted, thereby renewing natural resources.

All rights reserved.
No part of this publication may be copied, reproduced or transmitted in any form or by any means without written permission of the publishers or else under the terms of any licence permitting limited copying issued by the Irish Copyright Licensing Agency, The Writers' Centre, Parnell Square, Dublin 1.

Map extracts are reproduced by permission of Ordnance Survey, Ireland.

OUR DYNAMIC WORLD: THE SERIES

Ordinary Level Students use **TWO** textbooks:
* ✳ *Our Dynamic World 1* (plus optional workbook)
* ✳ *Our Dynamic World 2* **or** *Our Dynamic World 3*

Higher Level Students use **THREE** textbooks:
* ✳ *Our Dynamic World 1* (plus optional workbook)
* ✳ *Our Dynamic World 2* **or** *Our Dynamic World 3*
* ✳ *Our Dynamic World 4* **or** *Our Dynamic World 5*

CORE
All students must cover Book 1 (Workbook highly recommended)

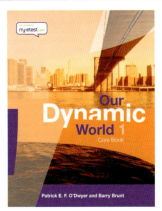

Book 1: covers the core sections of the syllabus which must be taken by all students

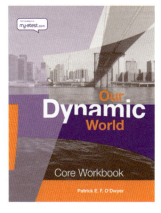

A workbook to accompany *Our Dynamic World 1*

ELECTIVES
All students must cover *either* Book 2 *or* Book 3

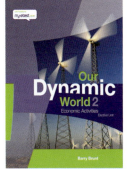  ⇐ **OR** ⇒

Book 2: Economic Activities – Elective Unit

Book 3: The Human Environment – Elective Unit

OPTIONS
Higher Level only students cover *either* Book 4 *or* Book 5

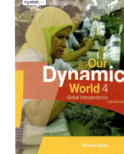

Higher level only

⇐ **OR** ⇒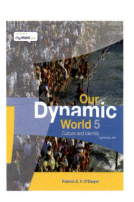

Book 4: Global Interdependence – Optional Unit

Book 5: Culture and Identity – Optional Unit

UNIVERSITY COLLEGE Library Cork

SYLLABUS OUTLINE

ELECTIVE UNIT: PATTERNS AND PROCESSES IN THE HUMAN ENVIRONMENT

	Content description	National settings	International settings
1	**Economic Development** Statement: Economic activities are unevenly distributed over the earth. (chapter 1) Students should study ● gross national product as a measure of economic development ● the human development index as a broad measure of development.	Ireland	Appropriate relevant European and global examples.
2	Statement: Levels of economic development show major spatial variations and can change over time. Levels of economic development evolve through the complex interaction of factors including physical, social, cultural, and political. (chapters 2–7) Students should study ● a case study from developed economies. This case study should include regions dominated by service and footloose industries, financial services and/or mass tourism regions. The case study should also, where appropriate, examine evidence of industrial decline ● a case study of a developing economy outlining the impact of colonialism, and adjustments to a global economy ● the global issues and a justice perspective relating to these patterns.	Ireland	Appropriate relevant European and global examples.

3	**The Global Economy**
	
	Statement:
	A single interdependent global economy has emerged with different areas having different roles. (chapters 8–13)
	
	Students should study
	● one multi-national company (MNC) to gain an understanding of its structure and organisation. For one product of the selected MNC students should be aware of the
	  – sourcing of raw materials and components
	  – location of basic processing units
	  – location of markets
	● patterns in world trade show that economies have become linked within a global framework. A trading pattern has emerged involving the USA, Europe, and the Pacific rim countries. These are the three key global economic areas. An international division of labour has emerged.
	
	Students should study how
	● basic processing units are widely spread
	● core and peripheral regions have developed
		An Irish based MNC	Appropriate relevant European and global examples.
			
			Show the global nature of linkages

ELECTIVE UNIT: PATTERNS AND PROCESSES IN THE HUMAN ENVIRONMENT

	Content description	National settings	International settings
	• some regions are excluded from world manufacturing activity • globalisation has impacted on world trade. Students should study one MNC to examine • the mobility of modern economic activities • how corporate strategies influence the opening and closure of branch plants • how product life cycle leads to changes in location • the future development of economic activities, teleservices, information technology, and e commerce.	Appropriate examples of MNCs based in Ireland.	

4	**Ireland and the European Union**
	
	Statement:
	Ireland as a member of the EU is part of a major trading bloc within the global economy. (chapters 14–18)
	
	Students should study
	• EU trading patterns within the single market and also external EU trade
	• Irish trading patterns with the EU and how the EU influences the Irish economy, for example:
	  – common agricultural policy
	  – common fisheries policy
	  – regional development funds
	  – social funding
		Appropriate relevant national examples.	Internal and external trading patterns

5	**Environmental Impact** Statement: **Economic activities have an environmental impact.** **(chapters 19–25)** Students should study • the use of renewable and non-renewable resources in the economy • the impact of the burning of fossil fuels and the use of alternative energy sources • environmental pollution at a local/national and global scale • sustainable economic development so as to control its environmental impact. Students should examine past experiences, future prospects and the necessity for environmental impact studies • conflicts that may develop between local and global economic interests and environmental interests. Students should be familiar with the issues relating to at least two examples.	National energy resources Smoke free zones. Patterns of production and consumption. National issues, the role of the EPA. Depletion of fish stocks, mining sites. Appropriate national examples e.g. Irish fish stocks, tourism and heritage.	Production and consumption of energy – appropriate European examples Acid rain – a European issue. Relevant global issues, e.g. global warming. Appropriate global examples.

CONTENTS

Economic Activities Have an Environmental Impact

Picture Credits

For permission to reproduce photographs and other material, the authors and publisher gratefully acknowledge the following:

ALAMY: 2L, 7R, 78, 92, 96, 124 all, 126, 127B, 145R © Alamy Images

PETER BARROW: 22, 24T, 26, 59, 68, 128, 137 © Peter Barrow Photography

CORBIS: 2R © Roger Ressmeyer; 6 © Paul W. Liebhardt; 7L © Robert Holmes; 9T © Alison Wright; 9B © Durand Patrick/Corbis Sygma; 10B © Jonathan Blair; 13T, 44B, 85T © David Turnley; 14 © Earl & Nazima Kowall; 15T, 48T, 109T © Michael S. Yamashita; 18 © Francoise de Mulder; 19 © Yang Liu; 32 © Jose Fuste Raga; 34T © Michael Maslan Historic Photographs; 34B © Hulton-Deutsch Collection; 36L © Carrion Carlos/Corbis Sygma; 36R © Szenes Jason; 37, 74 © John Van Hasselt; 42L © Tom Stewart; 42R © Bill Gentile; 44T © Joel Stettenheim; 47L © Julian Calder; 48B, 133 © Bettmann; 49 © Derek Cattani; 53T © Campbell William/Corbis Sygma; 83 © Langevin Jacques/Corbis Sygma; 85B © Matthew McVay; 94 © Ponti/Grazia Neri/Corbis Sygma; 95, 116, 160BL © Michael St. Maur Sheil; 105, 164T, 168TL © Yann Arthus-Bertrand; 111T © Peter Turnley; 114 © Contifoto/Corbis Sygma; 127T, 132R © Sally A. Morgan/Ecoscene; 140 © Martin Jones/Corbis; 146, 168TR © Tiziana and Gianni Baldizzone; 164B © Stephanie Maze; 165 © Jim Zuckerman; 166L © Collart Herve/Corbis Sygma; 166R © Owen Franken

IRISH IMAGE COLLECTION: 107, 109B, 111B, 113, 117, 151, 155, 160T, 160BR © Irish Image Collection

FINBARR O'CONNELL: 23B, 65TR, 65B, 91L, 99T, 149B © Finbarr O'Connell Photography

PANOS: 1L © Penny Tweedie; 15C, 90 © Irene Slegt; 15B, 79 © Crispin Hughes; 38 © Caroline Penn; 39 © Chris Stowers; 46L © Jon Spaull; 46R © Stefan Boness; 50B, 77T, 88 © Mark Henley; 53C © Trygve Bolstad; 72 © Morris Carpenter; 145L © Rob Huibers; 168B © Jeremy Hartley

PHOTOCALL IRELAND: 1R, 82, 91R, 123, 134, 152 © Photocall Ireland

REUTERS: 86 © Stringer/India

REX: 47R © Ashwin Gatha; 76 © Alex Segre; 142 © TDY; 143 © Sipa Press

SCIENCE PHOTO LIBRARY: 132L © Simon Fraser; 144 © Martin Bond

SKYSCAN: 13B © Kevin Dwyer; 31 © Henderyckx; 53B © LAPL; 71 © Aerophoto Schiphol

OTHERS: 3L © Peter Ginter; 3R © Peter Menzel; 10T, 51 courtesy of Daimler Chrysler; 23T © Neil Warner/warnerphoto.org; 24B © Simmons Aerofilm; 27 courtesy of McConnells Advertising/IDA Ireland; 30 © Harald Finster; 41B courtesy of Bewley's; 55 courtesy of Intel; 61 courtesy of Volkswagen; 65TL courtesy of Ford of Spain; 67 courtesy of the Buildings Dept, University of Limerick, taken by Press 22; 77B © Ward Scott; 97 courtesy of Shannon Development; 99B courtesy of Dublin Port Company; 101 courtesy of Liebherr, taken by Barry Murphy; 119T © Aer Rianta; 119B © Fennell Photography/courtesy of FÁS; 120 courtesy of City West; 136 courtesy of ESB; 138 © Garry O' Neill; 139 © Martyn Goddard; 141 courtesy of Galway City Council; 149T © courtesy of NRA; 150 © Harald M. Valderhaug; 153 courtesy of the author; 154 © R. Blakeman/Tara Mines; 158 courtesy of Save the Swilly;

CARTOONS: 20 both, 41T, 106 © Martyn Turner; 50T © Andy Singer/andysinger.com

The author and publisher have made every effort to trace all copyright holders, but if any has been inadvertently overlooked we would be pleased to make the necessary arrangements at the first opportunity.

SECTION 1 (CHAPTER 1)
ECONOMIC DEVELOPMENT

The concept of economic development involves more than levels of national and personal prosperity. It also includes a consideration of the quality of life of a population.

At the global level, major differences exist in terms of economic development.

This section introduces a number of key indicators that are used to measure and show differences in patterns of global economic development. Differences within the European Union and Ireland are also introduced:

● Chapter 1 Patterns of Economic Development

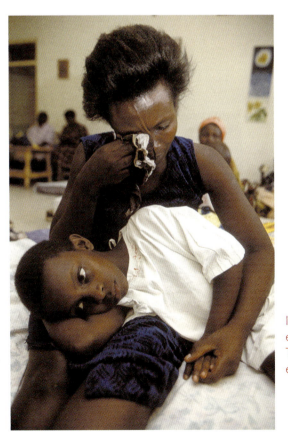

In sub-Saharan Africa, AIDS greatly reduces the life expectancy and quality of life for some 3 million people. They are 'worlds apart' from the lifestyles and prosperity enjoyed by people in places such as Temple Bar in Dublin.

CHAPTER 1
PATTERNS OF ECONOMIC DEVELOPMENT

KEY IDEA!

Economic activities and wealth are unevenly distributed over the earth.

THE MEANING OF ECONOMIC DEVELOPMENT

Economic development is a measurement of the prosperity of a region/country in terms of the strength of its economy and the quality of life of its population.

For many people, the term **economic development** is related simply to levels of prosperity. So, economic development is considered to be strong when people are well off, have secure, well-paid employment, and their future prospects look good.

Economic development is, however, a more complex process than one linked only to economic factors. It should also include the standard of living or **quality of life** of a population. This involves issues such as access to good medical care, education and a healthy diet. In effect, a country experiencing high levels of economic development has the wealth to ensure a good quality of life for most of its population.

Traditional farming in sub-Saharan Africa. Would this provide a high standard of living?

High-tech industry in Japan.

Domestic possessions of the Yadev family in India.

Domestic possessions of the Cavin family in the USA.

Class activity

Carefully study the above photographs and those on page 2. Use the images to explain the meaning of **economic development** and **quality of life**.

MEASUREMENT OF ECONOMIC DEVELOPMENT

Gross National Product

Economic development can be measured by a number of indicators such as:

● infant mortality rate

● life expectancy

● percentage employed in agriculture

● literacy rate.

However, the single indicator that is used most frequently to illustrate levels of economic development is the **gross national product per head for population** (GNP per person). It is usually measured in US dollars ($).

Since the price of goods and services differs greatly between countries, the purchasing power of a dollar can vary significantly. For example, in a country such as Ethiopia, one dollar will buy far more than in the United States. In order to take this into account, and to make international comparisons in income levels more effective, we use the term **purchasing power parity** (PPP). This converts a national income to its equivalent in the USA. We note this as **GNP per person (PPP)**. This is the preferred way to illustrate international differences in income.

> The GNP of a country is the total value of all output produced by that country's economic activities, including any net income from abroad. Dividing this value by total population provides the average GNP per person.

> The GNP per person in 2000 for Ethiopia was $100. When this is converted to PPP, its value in terms of purchasing power is increased to $668.

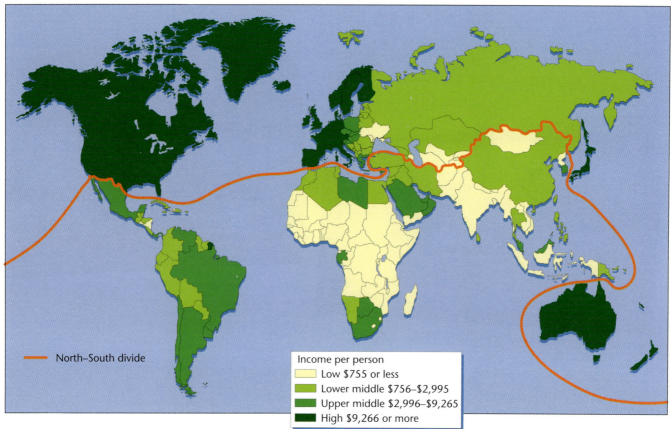

Fig. 1.1 Global map of GNP per person

Class activity

Study Figure 1.1 and answer the following:

1. Identify the *three* global regions of high income located in the northern hemisphere.
2. What global regions have the lowest levels of income per person?
3. Estimate how many times a person living in a high-income country is better off than someone in a low-income country.
4. Does GNP per person justify the use of a North–South dividing line to illustrate differences in levels of global development? Explain.

Human Development Index

Economic development is a complex process and involves a number of demographic and social conditions, as well as economic factors. As a result, most people think that no single indicator is adequate to illustrate different levels of economic development.

The United Nations has devised a way to measure the complexity of development through its **Human Development Index** (HDI). This combines three different indicators of development:

- life expectancy (demographic factor)
- GNP per person (PPP) (economic factor)
- adult literacy rates and enrolments in school (social factor).

By combining these three factors, the HDI is considered to be an effective measure of development. The index has a range of values from 0.0 to 1.0. The higher the value of the HDI, the more developed the national economy (Figure 1.2 and Table 1.1).

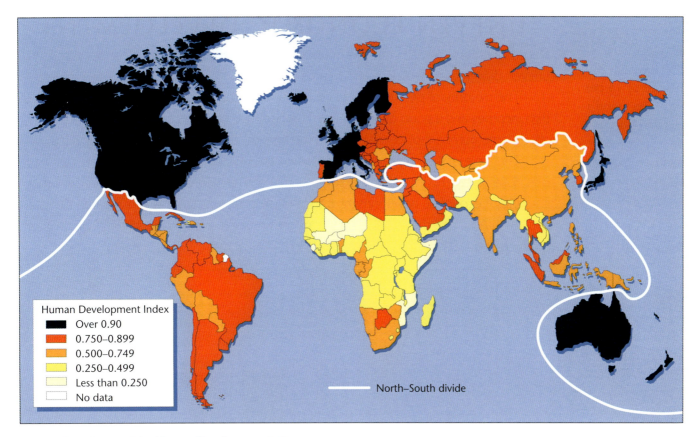

Fig. 1.2 Global map of the Human Development Index

Table 1.1 Sample indicators of development in selected countries

	GNP per person ($PPP)	Infant mortality Rate 1	Literacy Rate 2	Poverty Rate 3	HDI
USA	34142	7	100	–	0.939
Ireland	29866	6	100	–	0.925
India	2353	69	57	44	0.577
Ethiopia	668	117	39	31	0.327

1. per 1,000 live births.
2. percentage of relevant age group enrolled in secondary education.
3. percentage of population living on less than $1 a day.

Class activity

Study Figure 1.2 and Table 1.1 and answer the following:

1. What global regions have the lowest levels of HDI?
2. Why are the levels so low in these regions?
3. Which global regions have the highest levels of HDI?
4. Compare Figure 1.2 with Figure 1.1 Do you see a relationship between levels of HDI and GNP per person?
5. Use Table 1.1 to discuss differences in levels of economic development between Ireland, India and Ethiopia.

UNEVEN ECONOMIC DEVELOPMENT

For geographers, perhaps the most important characteristic of global development is its uneven distribution (Table 1.1 and Figures 1.1, 1.2).

The majority of people living in countries in the northern hemisphere have relatively high incomes and a good quality of life. This contrasts with the extreme poverty that affects large numbers of people living in the tropical zone and southern hemisphere.

The scale of the problem is illustrated by the numbers of people living on less than one dollar a day (Figure 1.3). This suggests that a huge effort will be necessary by richer countries if economic development is to take off in regions such as sub-Saharan Africa and south Asia, and their large populations are to begin to enjoy the benefits of a higher quality of life.

> Eighty per cent of global wealth is produced by 15 per cent of the world's population living in a small group of countries located mainly in the northern hemisphere.

> Approximately 20 per cent (1.2 billion) of the world's population lives on less than one dollar a day.

> The world's richest 1 per cent of people receive as much income as the poorest 57 per cent.

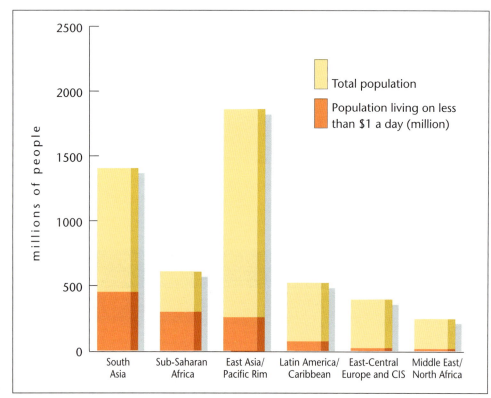

Fig. 1.3 Distribution of population living on less than US$1 a day

In India, an estimated 44 per cent of the population lives on less than one dollar a day. Is this a problem for the development of this country?

Class activity

Study Figure 1.3.

1. What global regions have the greatest number of people living in extreme poverty?
2. Estimate the percentage of people living on less than one dollar a day in Sub-Saharan Africa.
3. Suggest some reasons as to why so many people are living in extreme poverty in such regions.

UNEVEN PATTERNS OF AGRICULTURAL AND INDUSTRIAL ACTIVITIES

The wealth of a country is influenced strongly by its levels of:

- agricultural development
- industrial development.

The Role of Agriculture

Despite being recognised generally as a declining economic sector, especially in terms of job opportunities and wealth creation, agriculture has a powerful influence on patterns of development.

The role of agriculture is particularly important in influencing low levels of income and quality of life within less-developed countries (Figure 1.4). This is related to:

- Few alternative prospects for employment – large numbers of people are forced to remain in farming.
- Severe population pressure on a limited land area means that **subsistence farming** dominates most areas.
- Difficult environmental conditions, such as drought, low soil fertility and soil erosion, add to the problems of productive farming.
- Low levels of education make it difficult to introduce new farming skills and techniques.
- Poor access to markets.
- The price of cash crops, such as coffee and rubber, is unstable in world trade so profits are often low.

> Over 50 per cent of all workers in the world are engaged in agricultural activities.

Farming in Morocco, North Africa, and the Paris Basin in France. Describe the contrasting farming activities and environments, and explain which provide a higher standard of living.

7

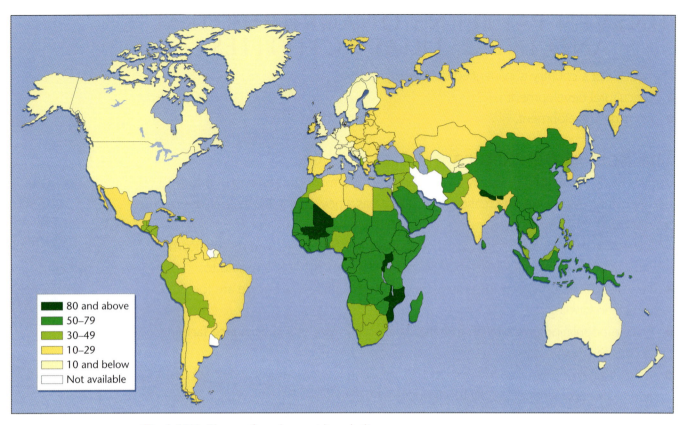

Fig. 1.4 World map of employment in agriculture

Fig. 1.5 World map of industrial production

Class activity
Study Figures 1.3 and 1.4.
1. Identify the global regions that have the lowest and highest dependency on agricultural employment.
2. Discuss why such a high dependency on agriculture in sub-Saharan Africa influences the region's low levels of development.
3. Identify the global regions that dominate industrial production.
4. Which continent shows the lowest levels of industrial production? Explain.
5. What general conclusions do you make when comparing the maps in Figures 1.3 and 1.4 with those in Figures 1.1 and 1.2.

The Role of Industry

Since the industrial revolution, the growth of industry has been recognised as a key sector for the promotion of economic development. However, not all areas of the world have been successful in attracting large-scale industrialisation (Figure 1.4).

Countries with high levels of industrial production generally benefit from high wages and standards of living. The workers' income, together with the requirements of industry for a range of services, such as finance, marketing and legal, also encourage the growth of a well-developed service sector. This further helps to raise levels of prosperity and the quality of life of a population.

Areas that show a well-developed industrial economy are usually linked to:
● early development of industry which gives a long-established tradition
● a large and prosperous home market
● well-developed transport and communication networks to give access to the growing world market
● availability of capital and services, such as research and development, which help improve productivity and competitiveness
● a well-educated and skilled labour force.

Employment in services has also grown strongly in less-developed countries. Here, however, the poor industrial base and low income levels for the majority of people usually mean that the quality and range of services are limited e.g. street vendors.

In the late twentieth century, a small number of formerly less-developed countries increased significantly their industrial economy. This was based on strong government support and a relatively cheap but flexible labour force. These are called newly industrialised countries (NICs). Examples include Singapore, South Korea and Mexico.

Industrial development in many less-developed countries often involves large numbers of females who work for low wages in sweatshops. Why do you think they are called sweatshops?

Large-scale industrial development and trade ports such as Le Havre in France create great prosperity.

UNEVEN DEVELOPMENT IN THE EUROPEAN UNION

Name the three
most developed
global regions.
Refer to
Figures 1.1 and 1.2.

Although the European Union (EU) is one of the three most developed global regions, considerable differences in levels of economic development exist within its boundaries. These differences occur between countries and within each of the member states (Figure 1.6).

At the EU level, a well-defined axis of urban-industrial development can be identified from Manchester to Milan. Increasingly, however, the core of prosperity is centred around the Alps in southern Germany and northern Italy. This contrasts with an extensive area of low income levels in its Mediterranean periphery.

Each country of the EU has its core and periphery, and significant differences in economic development exist between them. One of the best examples of this is between the well-developed north of Italy and the country's underdeveloped Mezzogiorno in the south.

Greater London, the richest region in the EU, is six times as prosperous as Iperios, the poorest region located in Greece.

Recall the meaning of core and peripheral regions.

Stuttgart in southern Germany forms part of the core of the EU. What evidence can be seen in the photograph which supports the fact that this is one of the richest cities in the EU?

The peripheral location and difficult environmental conditions, especially in the Apennines, continue to make the Mezzogiorno one of the least-developed regions in the EU. Refer to the photograph to justify this statement.

Class activity

Study Figure 1.6.

1. What countries/regions form the underdeveloped Mediterranean periphery of the EU? Justify your selection.
2. Identify the Manchester–Milan axis. Suggest reasons why this axis is so wealthy.
3. Suggest why eastern Germany experiences such low levels of prosperity compared to western Germany.
4. What is the dominant characteristic of member states that joined the EU in 2004? Will this be a problem for the EU?

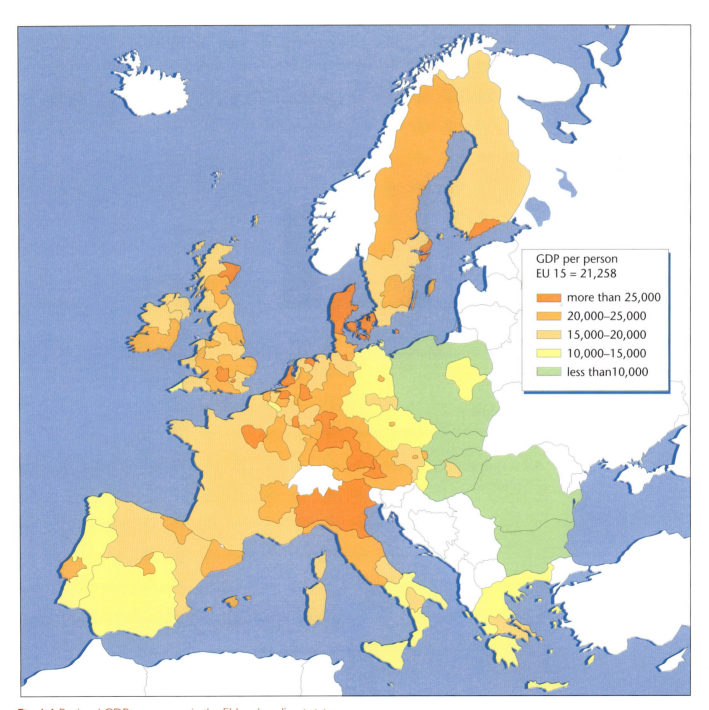

Fig. 1.6 Regional GDP per person in the EU and applicant states

REGIONAL INEQUALITIES IN IRELAND

Ireland has been viewed traditionally as part of the underdeveloped Atlantic periphery of the EU. In the 1990s, however, the country's levels of prosperity increased significantly under what became known as the **Celtic Tiger economy**. As a result, Ireland's GDP per person increased to rise above the EU average.

The new Irish prosperity was not, however, evenly distributed throughout the country. Effectively, it was the South and East region, and especially the Greater Dublin area, that gained most from the many high-tech industries and services that located and expanded in Ireland. As a result, the Border–Midlands–West (BMW) region remains underdeveloped, with income levels significantly below those of the South and East region, and the EU. In contrast, therefore, to the South and East region, which now has income levels above the EU average, the BMW has retained its Objective 1 status for Structural Funds from the EU.

What is meant by the term 'Celtic Tiger' economy?

What is meant by an 'Objective 1' region? See *Our Dynamic World 1*, p. 188.

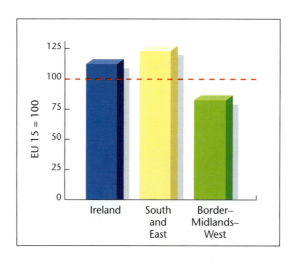

Fig. 1.7 GDP per person in Ireland's regions

Class activity
1. Name the least prosperous region in Ireland.
2. Which region has income levels above both the Irish and EU averages?
3. Suggest some reasons for the differences in prosperity within Ireland.
   (Refer to *Our Dynamic World 1*, Chapter 18).

TEST YOURSELF AT
my-etest.com

CHANGING PATTERNS OF ECONOMIC DEVELOPMENT

This section focuses on economic development as a process which shows considerable change over time. These changes in economic development are due to a complex interaction of factors, including physical, social, cultural and political.

Case studies from Ireland and Belgium are used to highlight changing patterns of economic development in the developed world, while the evolution of economic development in the less-developed world is reviewed through the impact of colonialism and decolonialisation. Finally, unequal economic development is illustrated from the point of view of justice or 'fair play'.

This section has six chapters:

- Chapter 2 – Changing Patterns of Global Economic Development
- Chapter 3 – Economic Development and Regional Change in Ireland
- Chapter 4 – Changing Patterns of Economic Development in Belgium
- Chapter 5 – Colonialism and Development
- Chapter 6 – Decolonisation and Adjustments to the World Economy
- Chapter 7 – Global Issues of Justice and Development

Poverty and civil wars cause millions of people to migrate in Africa and give rise to large refugee camps such as this one in Tanzania.

American companies, such as Intel, have been attracted to Ireland and have helped modernise the economy.

CHAPTER 2
CHANGING PATTERNS OF GLOBAL ECONOMIC DEVELOPMENT

KEY IDEA!

Economic development is not only unevenly distributed, but levels of development also change through time.

Economic development is a **process of change** that affects the make-up of a region's economy, together with its levels of prosperity.

Chapter 1 defined economic development as a measurement of a region's prosperity and quality of life. In this sense, the emphasis is on uneven patterns of development at one point in time.

Economic development, however, should not be seen as being static. It is, above all, **a dynamic process that results in significant changes in levels of development over time**.

Class activity
Discuss the differences in the definition of economic development as provided on this page with that found on page 2 in Chapter 1.

Generally, *five changes* are considered important for a region to become economically developed:

1. **Change in the structure of the economy**. This usually involves a shift from dominance of the primary sector (especially agriculture) to a greater role for industry and services. This creates a greater range of well-paid jobs and generates high-value manufactured goods and financial services.

2. **Changes in the use and levels of technology**. High productivity levels are often linked to the successful

Large areas of India continue to depend on agricultural practices that have changed little over the centuries. Describe the farming activity and suggest the crop being planted.

introduction and promotion of new technology. This can involve the use of new machinery, computers and other labour-saving practices. It helps regional/national economies to be more competitive in trade and increases prosperity.

3. **Changes in the forms of economic organisation**.
This involves a shift away from a relatively localised and simple way of life (subsistence), to one that demands more complex management. These tend to be larger-scale organisations that operate in the more competitive national/international market e.g. big businesses replacing small, family workshops, and large shopping centres which can cause the closure of small, community-based shops.

An automated assembly line using robots at a Toyota car plant in Japan. How many workers do you see? Is this a change from more traditional factories, suggesting higher productivity?

Economic development in less-developed countries often involves multinational companies. In what way does this help economic development?

4. **Changes to the economic well-being**. Economic development is expected to bring an improvement to the quality of life of a region's population. This involves improvements in services such as health care, education and housing. It is also expected to upgrade a region's transport systems and energy supplies to improve prospects for development.

5. **Changes in the volume and composition of trade**.
Economic development is influenced strongly by a country's ability to increase the volume and value of its trade. This is particularly important in terms of exports in order for a country to achieve a positive balance of trade. Through this, money flows into a country and this can be used to promote economic development. To raise the value of exports, it is important for a country to shift its dependency from relatively low-priced primary goods, such as foodstuffs and raw materials, to higher-valued industrial goods and services.

A school classroom in Mali in sub-Saharan Africa. How does this compare with your classroom? Why is education so important for economic development?

So, economic development should bring about a *decline* in poverty, unemployment and inequalities between different social groups and regions. It should also promote a *growth* in prosperity, quality of life and competitiveness.

What is meant by balance of trade? Why is a positive balance of trade important for economic development?

CHANGES TO GLOBAL PATTERNS OF ECONOMIC DEVELOPMENT

Since the end of the Second World War, the world economy has grown strongly. This has led to an increase in levels of development for most countries. Figure 2.1 illustrates changes in *four* key indicators that are used to measure education, health and income.

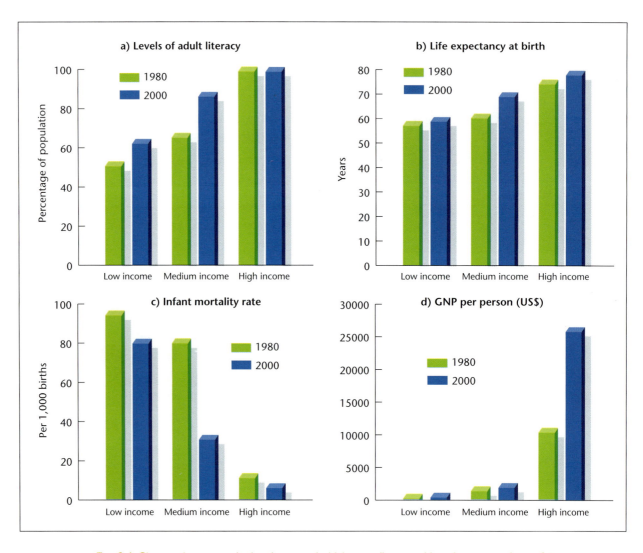

Fig. 2.1 Changes in economic development in high-, medium- and low-income regions of the world economy, 1980–2000

Class activity

Study Figure 2.1.

1. Describe the development trends shown by the indicators.
2. Apart from GNP per person, why does the high-income region show relatively small improvements in quality-of-life indicators?
3. Explain the link between income levels and quality-of-life indicators in the low-income region.

While most countries have shared to some extent in the process of economic development, the richer countries have shown a remarkable ability to capture most of the increases in wealth. In this way, the prosperity gap between the richest and poorest countries and regions of the world has increased over time. This can be illustrated especially through changes in income per person and poverty levels (Figures 2.2 and 2.3).

In other words, the rich have grown richer, while the poor have become poorer.

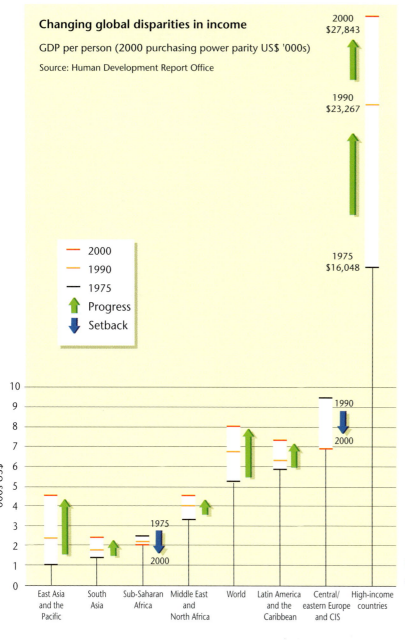

Fig. 2.2 Changing global disparities in income

Class activity

Study Figures 2.2 and 2.3.

1. Apart from central-eastern Europe, what other global region experienced a decline in income per person? Use the five factors outlined on pages 14–15 to suggest reasons for its decline.
2. What region showed greatest gains in prosperity?
3. What two world regions show an increasing proportion of world poverty?
4. What world region shows a significant fall in its share of world poverty?
5. Do you see any links between trends in poverty levels and changes in GNP per person?

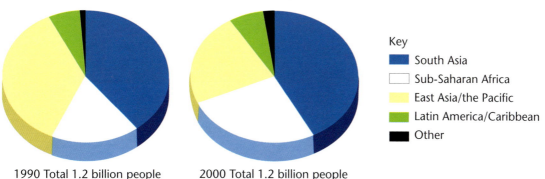

1990 Total 1.2 billion people   2000 Total 1.2 billion people

Key
- South Asia
- Sub-Saharan Africa
- East Asia/the Pacific
- Latin America/Caribbean
- Other

Fig. 2.3 Changes in regional distribution of world population living on less than US$1 a day

SUB-SAHARAN AFRICA IN THE 1990S – THE WORLD'S PROBLEM REGION

Fourteen of the 20 poorest countries in the world are located in sub-Saharan Africa. Most of these countries became poorer in the 1990s. This can be linked to:

- warfare and civil disturbances
- political instability which deters investment from multinational companies (MNCs)
- environmental problems e.g. droughts and increasing desertification
- failure to industrialise and reduce over-dependency on the primary sector for employment and export earnings
- health crisis linked to AIDs (almost 1 in 10 of the adult population has the virus).

Sub-Saharan Africa means the part of Africa that is south of the Sahara.

Many wars have broken out within and between countries in sub-Saharan Africa. Why do you think these have a negative impact on development?

Table 2.1 Selected development indicators for sub-Saharan Africa, 1990–2000

	1990	2000
Population (millions)	418	642
GNP (US$ billions)	161	308
GNP per person (US$)	385	480
Life expectancy	50	49
Infant mortality (per thousand all births)	104	107
Literacy rate (%)	45	62
Debt total (US$ billion)	63	230
Debt % GNP	39	75
Poverty (% population living on <$1 a day)	48	47

Class activity

Use the information in Table 2.1 to justify the belief that sub-Saharan Africa is the world's problem region.

Clearly, many countries in the less-developed world have experienced major difficulties in advancing their levels of economic development. Some countries, however, have shown more positive trends, linked mainly to a significant increase in their levels of industrial activity. These are termed **newly industrialising countries** (NICs).

What evidence is shown in this photograph to support the fact that Hong Kong has become a prosperous industrial and trade centre in south-east Asia?

Examples include the original four Asian Tiger economies of Hong Kong, South Korea, Singapore and Taiwan. These, together with other NICs that have emerged in east Asia and the Pacific, such as Indonesia and the Philippines, have helped raise levels of prosperity and lower levels of poverty in this world region (Figures 2.2 and 2.3).

Examples of NICs in other world regions include Brazil, Mexico and India.

At the start of the twenty-first century, levels of global inequality continued to increase. So, by 2000, the United Nations Report on Human Development described inequalities in income and living standards as having reached *grotesque proportions*. Clearly much work needs to be done if we are to experience a more balanced distribution in global economic development.

Class activity

Study the cartoons above.

1. Discuss what messages the cartoonist presents in trends in world development.
2. In what ways do you think inequalities in world development are grotesque?

TEST YOURSELF AT
my-etest.com

CHAPTER 3
ECONOMIC DEVELOPMENT AND REGIONAL CHANGE IN IRELAND

KEY IDEA!

East–west differences in economic development in Ireland have changed since political independence in 1922.

Regional development in Ireland is uneven. In particular, **a strong east–west divide between a more developed east and an underdeveloped west highlights the human geography of the country**. The nature and the extent of this divide have, however, changed since independence. This can be seen in terms of both manufacturing and service industries.

MANUFACTURING INDUSTRIES

Before independence in 1922, few industries were located in Ireland. Since then, however, the number and type of industries have grown substantially through four periods. These periods have had different impacts on the east and west of Ireland.

1922–61

In 1926, only 10 per cent of Ireland's workforce was employed in manufacturing. Over one-third of these were in Dublin.

To promote development, the Government adopted a **policy of protection**. This meant protecting the country's few industries from cheaper imports, and allowed Irish industries, such as clothing, shoes, food and drink, to grow.

These industries all showed a strong preference to locate in Dublin. Apart from some growth in other major cities, such as Cork, the rest of Ireland showed slow growth or decline in manufacturing employment (Figure 3.1).

From 1926 to 1961, the east–west divide increased as Dublin dominated industrial development.

1961–81

During this period, the policy of protection was replaced by one of free trade and attracting **multinational companies** (MNCs). This was successful and many branch plants of multinational companies located in Ireland.

Review Chapter 18 in *Our Dynamic World 1* to recall the major contrasts between the east and west of Ireland.

A policy of *protection* means placing tariffs on imports. This raises their costs and allows less efficient Irish industries to compete in the home market.

See Chapter 9 for a fuller explanation of branch plants and multinational corporations.

21

The Guinness Brewery has long been one of Dublin's major employers. Look at the photograph and suggest reasons for its successful location.

A footloose industry has the freedom to locate in a variety of locations.

These new branch plants were **footloose** industries. Their main concern was to find **low-cost locations** to mass-produce basic industrial goods. As a result, they showed a strong preference to locate their factories in rural areas and small towns. The result was large employment gains for western Ireland.

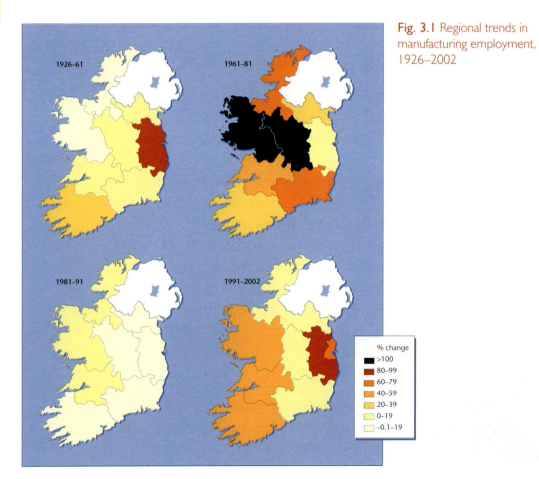

Fig. 3.1 Regional trends in manufacturing employment, 1926–2002

1926–61

1961–81

1981–91

1991–2002

% change
>100
80–99
60–79
40–59
20–39
0–19
-0.1–19

Class activity

Study Figure 3.1.

1. Which region showed the greatest growth from 1926–1961? Why?
2. How does the regional pattern for 1961–1981 differ from 1926–1961?
3. Describe how regional trends for the 1990s differ from the 1980s.
4. What two periods suggest a closing of the east–west gap for manufacturing employment? Explain.

In contrast, Dublin and other large cities, such as Cork, proved less attractive for MNCs. **From 1961–81, the manufacturing gap between east and west was therefore reduced** as footloose branch plants preferred to locate in rural Ireland.

Why would western Ireland be more attractive for branch plants than large urban areas?

Factories, such as this one located outside Charleston in Co. Mayo, illustrate the attraction of rural environments for footloose industries.

1981–91

As a result of economic recession in the 1980s:

- Fewer MNCs were attracted to Ireland, while many branch plants were closed or reduced their workforce.
- The closure of many traditional Irish industries caused large job losses.

Although growth of industrial employment in western Ireland slowed down compared to 1961–81, it performed better than the urban regions. Cork, and especially Dublin, experienced significant declines in their industrial workforce.

In the 1980s, therefore, the industrial divide between east and west was further reduced.

Old Ford/ Dunlop Site

Look at the photograph and suggest reasons why the Ford and Dunlop MNCs selected this site for their factories in Cork. Closure of both plants in 1984 resulted in the loss of 2,500 jobs. What effects would this have on the city?

23

Park West Industrial and Business Park in Dublin. Use the photograph to suggest reasons for its development at this location.

1991–2002

During this period of the Celtic Tiger economy, manufacturing employment grew strongly. This relates to a new wave of investment by MNCs.

Many of the new industries attracted to Ireland use modern technology to produce a range of high-value goods and services. These **high-tech industries** are also **footloose**, but are influenced by different locational factors than branch plants. These include:

- well-educated workers
- proximity to third-level education institutions
- access to high-quality transport and communications
- well-developed services.

Class activity
1. Would these location factors favour the east or west of Ireland? (Refer to Figure 3.1.) Explain your answer.
2. Did this result in a closing or widening of the industrial divide between the east and west?

Apart from the area centred on Galway, the rural west of Ireland has few attractions for high-tech industries.

SERVICE INDUSTRIES

1922–61

Service industries show a strong preference to locate in large urban centres. As a result, service employment in Ireland focused on the more urbanised eastern region. Dublin, in particular, dominated the country's service sector.

The **west of Ireland**, therefore, not only failed to attract manufacturing industries during this period, but also **failed to develop an effective services sector**. This widened the east–west development gap.

Dublin is the centre of government and most key services. This dominance has restricted the growth of strong regional centres.

1961–91

Employment in services grew rapidly in this period due to:

- increasing population
- industrial development
- increasing income levels.

Fig. 3.2 Regional trends in services employment in Ireland, 1961–2002

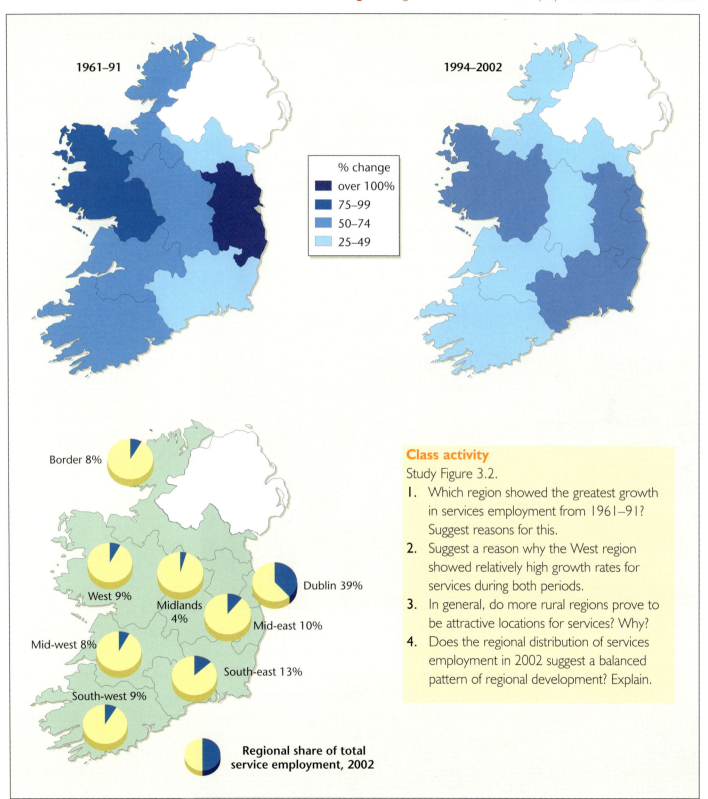

Class activity

Study Figure 3.2.

1. Which region showed the greatest growth in services employment from 1961–91? Suggest reasons for this.

2. Suggest a reason why the West region showed relatively high growth rates for services during both periods.

3. In general, do more rural regions prove to be attractive locations for services? Why?

4. Does the regional distribution of services employment in 2002 suggest a balanced pattern of regional development? Explain.

Almost 60 per cent of all new service jobs created in Ireland were located in the Eastern region

Although services employment increased in all regions, the East region, centred in Dublin, benefitted the most (Figure 3.2). In addition, Dublin and other key cities, such as Cork, Galway and Limerick, attracted most of the high-value office employment, e.g. banking and financial services. *Dublin*, above all, gained from this **centralisation of services**, and increased its role as the **dominant decision-making centre in Ireland**.

In contrast to manufacturing, therefore, **trends in services did not result in a narrowing of the development gap between the east and west of Ireland. Large areas of the west of Ireland depend on Dublin for many high-value services.**

1991–2002

All regions benefited from a strong growth in services (Figure 3.2). The high-value and well paid office employment continues, however, to prefer locations in major cities, especially Dublin. By 2002, almost 50 per cent of Ireland's total service employment was located in Dublin and the adjoining Mid-East region (Figure 3.2).

During the 1990s, a new factor emerged to influence the regional pattern of services. This was the inward investment of *footloose international service industries*. One element of this has been the relocation to Ireland of routine *back-office functions* of large, international companies.

The city of Galway, with its attractive urban environment, university and growing number of high-tech industries, has emerged as a major growth centre for services in the west of Ireland.

Back-office functions involve routine work and require little face-to-face contact, e.g. processing insurance claims and ticket reservations.

The vast majority of these back offices are located in Dublin because of its large pool of educated workers and excellent telecommunication systems.

In addition, therefore, to its national dominance, **Dublin has gained a new international role within the global economy**. This emphasises the historical development gap between the east and west in Ireland. It, furthermore, suggests that the gap may widen in the future.

Case study:

Two good examples of the rapid growth and centralisation of services in Dublin are call centres and the International Financial Services Centre (IFSC).

Call centres

Since the 1990s, a growing number of international call centres have been attracted to Ireland. Strong government support, high-quality telecommunications and a cheap but well-educated workforce were vital in this development (Figure 3.3).

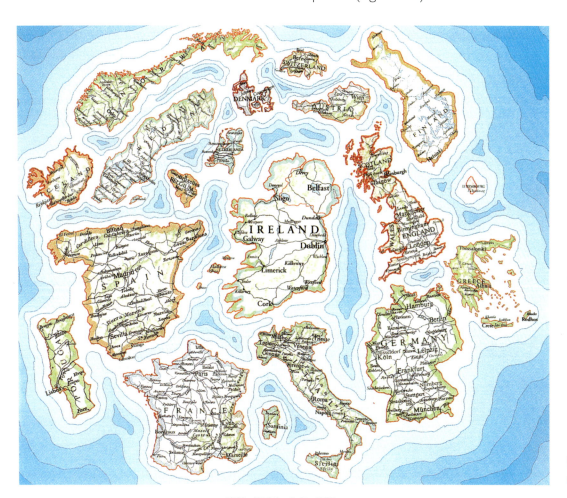

IRELAND.
THE CALL CENTRE OF EUROPE.

Many large European and American companies have recognised the value of setting up a pan-European call centre to serve each of their international markets.

If you are looking for the best location for your call centre, look no further than Ireland; thanks to our advanced telecommunications technology, no other country is closer to the heart of Europe - yet no other country has more competitive call rates.

Ireland can also offer a well educated, multilingual and flexible workforce at a lower cost. Add in a substantial tax benefit and you have the most effective Call Centre in Europe.

If you want to find out how you can join companies such as **ITT Sheraton**, **Best Western**, **The Merchants Group**, **Global Res** and **Dell** in making the most of Ireland's telecommunications advantage, give us a call.

HEAD OFFICE
Ireland
Wilton Park House, Wilton Place, Dublin 2.
Tel: +353 1 603 4000
Fax: +353 1 603 4040

United Kingdom
Ireland House, 150 New Bond Street, London W1Y 9FE.
Tel: (171) 629 5941
Fax: (171) 629 4270

email: idaireland@ida.ie   web: http://www.idaireland.com

THE CALL CENTRE OF EUROPE

A call centre is a central location from which services, such as technical support and sales, are provided to a dispersed customer base through the use of the telephone.

Ireland has the cheapest rates in Europe for international freephone calls. Why do you think this is important?

Fig. 3.3 Ireland: the call centre of Europe

Class activity
What do you think this Industrial Development Agency (IDA) advert is suggesting in order to attract inward investment of call centres?

27

By 2000, over 50 call centres had located in Ireland, employing over 6,000 people. **Over 90 per cent of all call centres are located in Dublin.**

Few call centres locate outside Dublin. This is due to **concerns over access to enough qualified workers, especially with language skills**, as well as less-developed services in smaller towns.

IFSC

In 1988, the IFSC was opened. This was the centrepiece of a major scheme aimed at revitalising a large area of under-utilised/derelict land along Dublin's inner quays.

By providing generous financial incentives, the plan was to attract footloose, international financial services to Dublin. The scheme has been remarkably successful. Dublin is now an important financial centre within the EU. This adds further to its international image to attract more high-quality service employment.

By 2002, over 430 companies employed more than 10,700 people in the IFSC.

TEST YOURSELF AT
my-etest.com

CHAPTER 4
CHANGING PATTERNS OF ECONOMIC DEVELOPMENT IN BELGIUM

KEY IDEA!

The core region of Belgium's economy has changed from Wallonia to Flanders.

Since the 1950s, Belgium's core region for economic development has shifted from Wallonia in the south, to Flanders in the north (Figure 4.1). This is due to **three** processes which have affected each region differently:

- the decline of mining and heavy industries
- the changing locational preferences of modern, footloose industries
- the increasing importance of services as a growth sector.

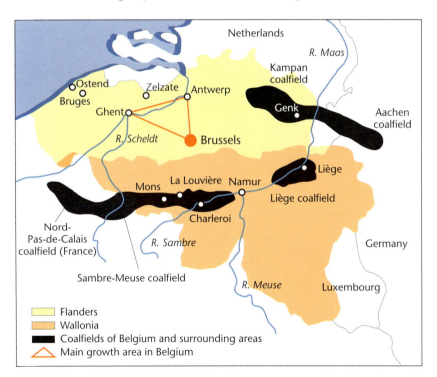

Fig. 4.1 Belgium, Flanders and Wallonia

Class activity

Study Figure 4.1.
1. Which region possessed the major Belgian coalfields?
2. In which region is the present growth area located?
3. What three towns form Belgium's growth triangle?

Wallonia: the historic core region

Belgium's main coalfields are located in Wallonia. This led to the large-scale development of mining and heavy industries throughout the Sambre Meuse Valley and around Liege. The coalfields enjoyed great prosperity from about 1800 to the 1950s.

In 1953, there were 123 coalmines in Wallonia, employing 120,000 miners. The last mine closed in 1984.

29

Since the 1950s, exhausted coal seams, falling productivity and rising costs meant that the region's coal and heavy industries could not compete against cheaper imports. For communities dependent on these industries, closure of the mines and decline of the steel industry resulted in large-scale unemployment.

Remember the development and decline of the Sambre Meuse in *Our Dynamic World 1*, Chapter 16.

The long tradition of mining and heavy industries in the Sambre Meuse valley had a major impact on its environment. What evidence of this impact can you see in the Charleroi region, as shown in this photo. Why is financial support from the EU and Belgium so important to help modernise this region's economy?

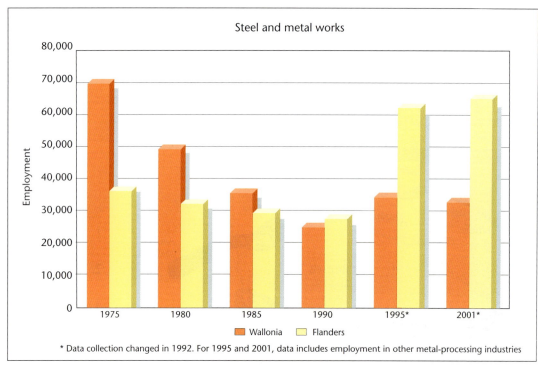

Steel and metal works

* Data collection changed in 1992. For 1995 and 2001, data includes employment in other metal-processing industries

Fig.4.2 Employment trends in the steel industry in Wallonia and Flanders

Wallonia has become a depressed region at the heart of the EU. Without Belgian and EU funding, decline would have been greater. Support was given to:
● modernise the region's declining steel industry to protect some jobs (Figure 4.2)
● retrain workers, improve transport systems and clean-up the despoiled environment
● attract new industries and services.

Class activity
Study Figure 4.2.
1. What was the peak year for employment in Wallonia's steel industry?
2. Describe the employment trend in Wallonia's steel industry.
3. How does this compare with the trend in Flanders?
4. Which is now the most important steel and metal-processing region in Belgium? Why?

In spite of this, industrial employment has been more than halved since 1975, while the depressed regional economy has been less attractive than Flanders for high-value services (Figure 4.3). One important result has been a large and continuous movement of people from Wallonia to Flanders and Brussels. This further depresses the economic and cultural position of Wallonia.

After 150 years of dominance, Wallonia's core status has been replaced by Flanders.

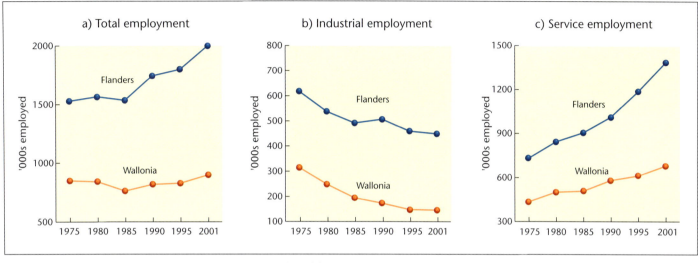

Fig.4.3 Employment trends for Flanders and Wallonia, 1975–2001

Antwerp is the third largest port in the EU and has been a critical factor in attracting a large amount of industrial investment to Flanders.

Class activity

Study the graphs shown above in Figure 4.3.
1. Which region shows the greatest rate of decline in industrial employment? Briefly explain.
2. Describe the regional trends for services employment. Suggest reasons why Flanders grew more strongly.
3. What are the main differences in total employment trends for Flanders and Wallonia? Explain.

Flanders: Belgium's modern core region

Before the Second World War, small traditional industries in Flanders could not compete with the large-scale factories of Wallonia. Many were closed and large numbers of Flemish people were forced to migrate to the prosperous coalfields of Wallonia.

Since the 1960s, however, Flanders has experienced significant economic growth. This is due to:

- its central location and excellent transport networks to access major markets in Belgium and the EU
- its access to Antwerp, Europe's third-largest port and the most important industrial centre in Flanders
- its attractive environment and many historic towns as locations for high-quality service sectors
- its growing, skilled and adaptable workforce
- the decision to open a major new steelworks at Zelzate, which has made Flanders the main steel-producing region in Belgium (Figures 4.1 and 4.2).

These factors have attracted large numbers of modern, footloose industries to Flanders. So, although industrial employment has declined, its manufacturing base is now made up mainly of growth sectors such as electronics and pharmaceuticals. In addition, Flanders has been remarkably successful in attracting high-quality service jobs, and especially those in business and financial services. This suggests that **prospects for continued growth and prosperity in Flanders are strong, and it will remain a core region of the EU**.

By 2001, Flanders had three times as many industrial jobs as in Wallonia.

Between 1975 and 2001, 90 per cent of the net gain in total employment for both Wallonia and Flanders was located in Flanders.

The historic city of Ghent forms with Antwerp and Brussels Belgium's triangle of growth. Compare this photograph with that on page 30. Why is Ghent more attractive for modern service industries and tourism?

TEST YOURSELF AT
my-etest.com

CHAPTER 5
COLONIALISM AND DEVELOPMENT

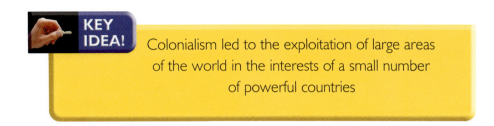

KEY IDEA!

Colonialism led to the exploitation of large areas of the world in the interests of a small number of powerful countries

Historically, *colonialism* is linked to a small number of European states which extended their economic and political control over large areas of the world (Figure 5.1). In this way, **colonialism had a major impact in changing the patterns of world development**.

Name some colonial powers in addition to Britain.

The Objectives of Colonialism

1. To control and exploit the **raw materials and food produce** of colonies. These were vital to supply the growing economies and populations of colonial powers.
2. Colonies would provide **markets** for goods manufactured by the colonial power.
3. Control over large areas of the world and the growing trade patterns linked to colonialism would increase the **political importance and wealth** of colonial powers.

British Empire

Fig. 5.1 Britain's colonial empire in 1914

At this time, the British Empire controlled 25 per cent of the world's population and land area. Why was it described as an empire 'on which the sun never sets'?

33

The Impacts of Colonialism

1. Most colonies changed from being generally self-sufficient to **specialising in a narrow range of primary products**. For example, local food crops were replaced by *plantation farming*. So, cash crops, such as bananas, coffee, rubber, were exported to meet the needs of the colonial power.

Field workers at a large tea plantation in India with baskets filled with harvested tea leaves (c.1900). Why did plantation farming result in the underdevelopment of colonies?

2. Before colonialism, countries now considered to be underdeveloped, dominated global manufacturing output (Table 5.1). Under colonialism, however, their **domestic industries were run down**. This was to remove competition from industrial goods made in countries like Britain which were exported to their colonies. By the First World War, manufacturing activities in the colonies had almost ceased. *Colonialism, therefore, contributed to major changes in patterns of global manufacturing.*

3. Colonialism created **a new and dependent pattern in world trade**. Colonies *specialised in producing primary goods for export* to colonial powers. In contrast, countries like Britain specialised in industrial production and exported these higher-valued goods to their colonies. This dependent relationship gave rise to an **international division of labour** (see Chapter 13).

4. *New transport networks (especially railways) and port cities* were developed in the colonies. The railways were to bring primary products to key port cities, such as Calcutta and Bombay in India, for export to their colonial power (see Chapter 21 *Our Dynamic World 1*). The rest of the colony remained underdeveloped.

British sailing ships crowd into Calcutta Harbour (c.1860). What role did such ports play to allow colonial powers to dominate these colonies?

Table 5.1 Changing patterns of world manufacturing output (%)

	1750	1830	1913
Developed market economies	27	40	93
e.g. Britain	2	10	14
USA	–	2	3
Underdeveloped economies	73	60	7
e.g. India	25	18	1

Class activity

Study Table 5.1.

1. What part of the world economy dominated manufacturing before the industrial revolution?
2. In what ways do the 1913 patterns differ from 1750?
3. Suggest reasons for the different trends for Britain and India.

5. Colonialism is **an exploitative process** by which colonial powers dominate their colonies. This increases the wealth of the colonial power, but results in *the underdevelopment of the colonies*.

Case Study: India – a changing colonial economy

Prior to independence in 1947, India had long been exploited as a British colony. Such was its importance for Britain's empire that it was known as 'the jewel in the British Crown'.

Until the early 1800s, India's large population and raw material base supported a variety of important craft industries, and especially textiles. These were all run down to allow British goods to gain access to a large, new market. So, as India's population increased, its industrial economy collapsed under colonialism (Table 5.1).

India specialised in exporting primary commodities such as cotton, tea, jute and spices to Britain. Taxation of the large population also added greatly to Britain's wealth. Key port cities such as Bombay, Calcutta and Madras developed under colonialism, although most of India remained underdeveloped and dependent upon a poor, subsistence economy.

Do you think these impacts of colonialism can be applied to Ireland, which was a colony of Britain until 1921?

'[British rule] has impoverished the people [of India] by a system of progressive exploitation . . . It has reduced us politically to serfdom.'
M. Gandhi, 1930

Why did India's port cities prosper while other areas did not?

TEST YOURSELF AT
my-etest.com

CHAPTER 6
DECOLONISATION AND ADJUSTMENTS TO THE WORLD ECONOMY

KEY IDEA!

Despite gaining political independence after the Second World War, most former colonies remain underdeveloped and economically dependent on core economies.

After the Second World War, most colonies gained political independence. This is a process called **decolonisation**. To help newly independent states adjust to the new world economy, a number of strategies have been used. These include:

- borrowing in order to finance development
- attracting new industries
- trade policies.

1. Borrowing

Interest repayments on debt amounts to over $260 billion. This is a huge loss of much-needed money for developing countries.

Colonialism left most colonies with little capital for development. Borrowing large sums of money from developed countries to help modernise their infrastructures, such as transport and energy systems, seemed an easy option. However, because of high interest rates, developing countries have been unable to pay back their loans. This is called the *Debt Crisis* (Table 6.1).

To help meet repayments, many developing countries have reduced their spending on important social programmes, such as health care and education. The result has been a decline in the quality or life for many people in developing countries.

So, *large scale borrowing has increased further their dependency on core economics.*

Bono of U2 with a petition he took to the United Nations in 2000 calling for the cancellation of the debts of the poorest countries to allow them to have a new start in the new millennium. Do you think this is a practical solution?

Major street demonstrations in Argentina to protest the financial constraints imposed by the government to help repay the country's debt crisis.

Table 6.1 Examples of the scale of world debt, 2000

	Total debt ($ billion)	Debt per person ($)	GDP per person ($)	Interest repayment as a % total export value
Sub-Saharan Africa	118.0	179	470	10
Ethiopia	5.5	85	100	14
Sierra Leone	1.3	255	130	48
Zambia	6.9	692	300	19
Central and South America	774.4	920	3,670	39
Argentina	146.2	3,950	7,460	71
Brazil	237.9	1,400	3,580	91
India	83.6	82	450	13

Class activity

Study the information in Table 6.1.

1. What are the total debts of sub-Saharan Africa and Central and South America?
2. Compare GDP per person and amount of debt per person in the countries of both global regions. Why is this a problem?
3. What do the figures in the final column suggest for the ability of a country, such as Brazil, to repay its debts?
4. Do you think India has a debt problem?

South Korea is a newly industrialising country that has developed rapidly. It now produces a wide range of industrial goods, and includes Hyundai Heavy Industries Co. which is the world's largest ship builder.

2. Attraction of New Industrial Development

From the 1960s, large multinational companies (MNCs) emerged as a powerful force for industrial development. As part of their plans to serve the growing and more competitive world market, MNCs began to invest in low-cost locations in developing countries.

Branch plants of MNCs therefore became an important new element in the economies of some developing countries. These rely on low-cost labour to produce basic goods which require few skills. This led to a *New International Division of Labour* (see pages 84–85). It also allowed developing countries to increase their share of world manufacturing output (see Figure 9.3 on page 58).

Comparatively few developing countries, however, have created a strong industrial sector. These are called **Newly Industrialising Countries (NICs)**. The vast majority of developing countries remain underdeveloped and continue to depend on the export of primary products (Figure 6.1). *Sub-Saharan Africa*, in particular, has found great difficulty in attracting new industries, due to wars, corruption and poverty.

For a fuller account of MNCs in global development, see Chapter 9.

Eighty per cent of all manufactured exports from developing countries are sourced in only ten NICs.

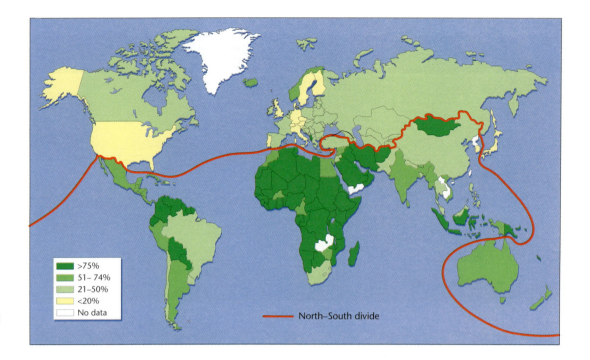

Fig. 6.1 Percentage share of exports held by primary goods

Legend:
- >75%
- 51–74%
- 21–50%
- <20%
- No data

—— North–South divide

Class activity

Study Figure 6.1.

1. Which world region is most dependent on the export of primary goods? Suggest reasons why.
2. Why is dependency on the export of primary goods so low in developed countries?
3. What level of dependency does India have on the export of primary goods? Do you expect this to fall?

3. Trade Policies

On gaining independence, many developing countries placed taxes, called tariffs, on imported goods. This was to *protect* home industries from cheaper imports. Today, however, **free trade** dominates the global economy.

Free trade removes restrictions for international trade and *encourages the export of goods for which a country has a comparative advantage.* This has helped increase trade for most developing countries. Their comparative advantage, however, remains in the production of primary goods and low-value, basic industrial products.

After some fifty years of political independence, most developing countries continue to depend on core economies. This favours rich, core economies and is termed **neo-colonialism**.

Picking coffee beans in Uganda. Many developing countries remain highly dependent on exporting primary products such as coffee. Why is this a problem for their economic development?

The modern face of development is seen in this high-tech manufacturing plant at Bangalore in India. Does this suggest that India is showing signs of adjusting well to the new world economy?

Case Study: Post-Independent India

Since independence, India has adopted elements of all three strategies to help adjust to the new world economy.

Until 1990, India followed a policy of *protection*. Its large population and resource base provided both a strong home market and the inputs needed for industrial development, e.g. textiles, steel, food processing. Since 1990, however, it has adopted *free trade*, as the country looked to benefit more from expanding world trade. This has been successful.

India also *borrowed wisely* and was not faced with a debt crisis. Investments were made to modernise the economy and upgrade infrastructure and education. In addition:

- *Many MNCs* have been attracted by its large market and low cost, but well-educated, workforce.
- Large amounts of high-tech goods and services are now produced in India.
- The value of industrial exports has increased significantly, and India is classed as an NIC.
- As exports of higher-value industrial goods and services increased, India's dependency on exports of primary goods decreased. (Why should this help promote development in India?)

TEST YOURSELF AT
my-etest.com

CHAPTER 7
GLOBAL ISSUES OF JUSTICE AND DEVELOPMENT

 **KEY IDEA!**

Large numbers of people have failed to benefit equally from the growing world economy.

Previous chapters have shown that global development is uneven. Developing countries, in particular, have generally failed to adjust successfully to new patterns of world trade. As a result, large numbers of the world's population feel a deep sense of injustice in that they have not been treated fairly by global development. This chapter introduces *three* areas of injustice.

- fair trade
- health
- gender discrimination.

1. Fair Trade

Many developing countries continue to depend mainly on the export of primary commodities, such as coffee and copper. This trade, however, is controlled by powerful MNCs which often work against the interests of developing countries.

- **Commodity prices** for goods such as coffee and copper have generally fallen since 1980 (Figure 7.1; Table 7.1). As a result, developing countries have to export more of a commodity simply to maintain their income levels. This can, however, flood the market and cause prices to drop even further. *Consumers in developed countries benefit from this, while producers suffer a major decline in incomes and standards of living.*

- The **terms of trade** have added to the problems of developing countries. While prices of primary commodities have declined, the prices of manufactured goods exported from developed countries have increased significantly. This is unfair, and is seen as an exploitation of populations in the developing world.

In 1972, Uganda had to sell 6 tons of cotton to import a truck made in Europe. By 2002, the import of a similar truck needed 35 tons of cotton.

40

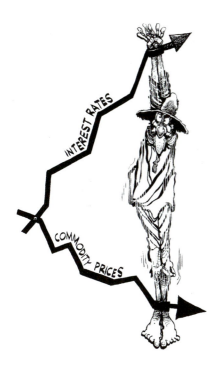

Table 7.1 Price changes for a sample of primary commodities, 1970–2000

	1970	1980	1990	2001
Cocoa (cents/kg)	240	230	127	111
Coffee (cents/kg)	300	412	118	63
Copper ($/tonne)	5,038	2,770	2,601	1,633
Cotton (cents/kg)	225	260	182	109
Rubber (cents/kg)	145	181	86	62

Experience of developing countries being torn between high interest rates on debt repayments and falling commodity prices.

Class activity

Study Figure 7.1 and Table 7.1.

1. What are the dominant price trends for primary commodities?
2. Contrast these trends with those for interest rates on loans made to developing countries.
3. What message does the cartoon suggest for producers of primary commodities?

● Producers of primary commodities **receive only a small proportion of the final value of a manufactured product**. Most of the profits are made by companies, which process the raw material inputs, and by retailers.

Fair trade organisations have been set up to try to ensure fairer prices and better working conditions for producers.

What does the Fairtrade symbol mean? Would you support such products, even if their price was slightly higher than competitor brands? Explain.

World trade is far from being fair. It has generally worked against the interests of most people in developing countries. This sense of economic injustice has to be addressed to achieve a fairer world economy.

41

2. Health: a matter of life and death

A fundamental human right is the right to life. Central to this is good medical care, which can help cure sickness, improve quality of life and increase life expectancy. This, however, is not equally available and results in major differences in health-related issues between populations in developed and developing countries (Figure 7.2).

Average life expectancy in Guatemala is 65 years compared to 77 in the UK.

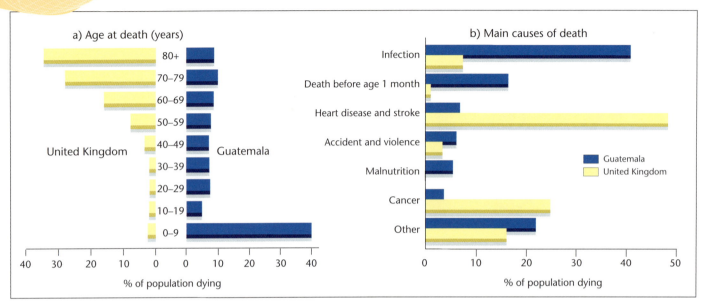

Fig. 7.1 Ages and causes of death in the UK and Guatemala

Class activity

Study the information provided in Figure 7.2.

1. What age group shows highest death rates in Guatemala? Compare this with the UK.
2. Contrast the main causes of death in Guatemala with those in the UK?
3. What do these trends tell you about health care and quality of life in both countries?
4. Does this suggest equal social justice for both populations?

People in developed countries, such as Britain, are generally well fed.

Many people in less-developed countries, such as these poverty-stricken children in Guatemala, are undernourished and are forced to scavenge for food from garbage dumps.

Class activity

What do these two images tell you about the health of people in Britain and Guatemala?

AIDS: a disease of the poor?

Since it first appeared in the late 1970s, over 25 million people have died from AIDS/HIV. Although considered to be a disease that can occur throughout the world, 90 per cent of all cases are in the developing world (Figure 7.3).

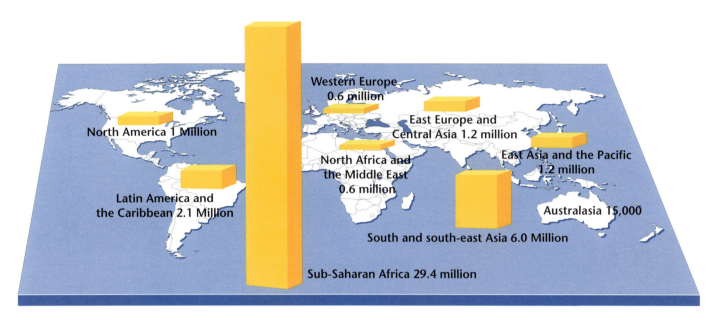

Fig. 7.2 Estimated world distribution map of people infected with AIDS/HIV

Class activity

Study Figure 7.3.

1. Which world region dominates the AIDS problem?
2. Estimate the number of AIDS cases in the developed world regions. What does this tell you about the distribution of the disease?
3. Discuss the ways in which AIDS can affect development in sub-Saharan Africa.
4. Do you think the developed world should take more responsibility in helping to reduce the AIDS crisis?

Sub-Saharan Africa: the AIDS 'hotspot'

Of the 42 million people in the world living with AIDS/HIV, 70 per cent are in sub-Saharan Africa. In 2002, 3.5 million new cases were reported in this region. All 20 countries of the world with the highest levels of AIDS/HIV cases are located in sub-Saharan Africa.

A nurse assists a patient at an AIDS hospital in Zambia. Throughout sub-Saharan Africa, the AIDS crisis has put a huge strain on health services, and has caused life expectancy to fall and infant mortality rates to rise.

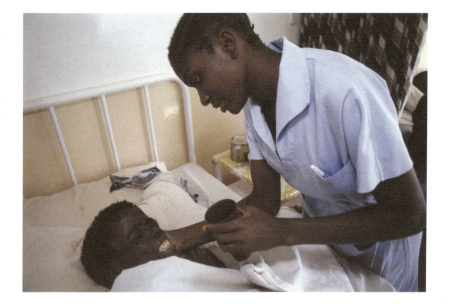

Given the scale of the AIDS crisis, especially in sub-Saharan Africa, it is both unfair and impractical to expect the poorest of world regions to deal with this health problem. Developed countries must commit more resources to such regions to combat AIDS. This would help reduce the huge inequalities in health care experienced by so many people in sub-Saharan Africa.

Access to low-cost and effective medicine would be a major help to combat AIDS in sub-Saharan Africa.

'AIDS today in Africa is claiming more lives than the sum total of all wars, famines, and floods, and the ravages of such deadly diseases as malaria … AIDS is clearly a disaster, effectively wiping out the development gains for the past decades and sabotaging the future … History will judge us harshly if we fail to act now, and right now.'

Nelson Mandela, 2000

Famine is also a massive problem for large areas of sub-Saharan Africa. In what ways do famine and AIDS affect economic development?

Class activity
Discuss the extent to which you feel Nelson Mandela is correct in his assessment of the AIDS crisis.

A mother with her starving child in Somalia. Here, war, drought and poverty combine to cause severe famines and the deaths of large numbers of people.

3. Discrimination: an issue of gender

Injustice in the world extends to gender. Despite the ideal that all people are equal, women have long been discriminated against in favour of men. This has important implications for development and cannot be justified for half of the world's population.

Women face *discrimination* in various ways:

- In some societies, the *law discriminates against women*. For example, in certain Moslem societies women do not have equal rights in marriage and face restrictions in working outside of the home.
- Although the situation is improving, males *generally are given easier access to education* than females. This reduces the opportunities for females to improve their employment prospects and quality of life.

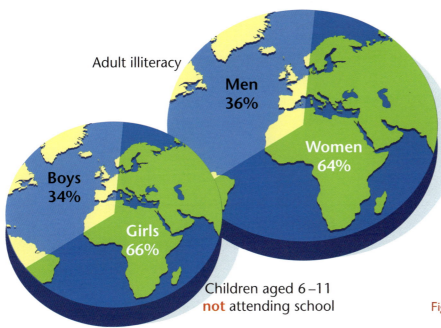

Adult illiteracy
Men 36%
Women 64%
Boys 34%
Girls 66%
Children aged 6–11 **not** attending school

Fig. 7.3 The gender gap in education

Class activity
Study Figure 7.4.
1. Do women or men account for most of the world's illiterate population?
2. Do girls or boys have a better chance of becoming educated?
3. Why should these figures be a cause of concern?

- The *preference for male children/heirs*, especially in traditional rural societies, is strong. This has resulted in selective abortion, or even the killing of girl babies.
- *Arranged marriages* often force young women into unsatisfactory relationships.
- Women are identified as *a cheap source of workers*, especially in developing countries.

Refer to the description of women workers in developing countries on page 88.

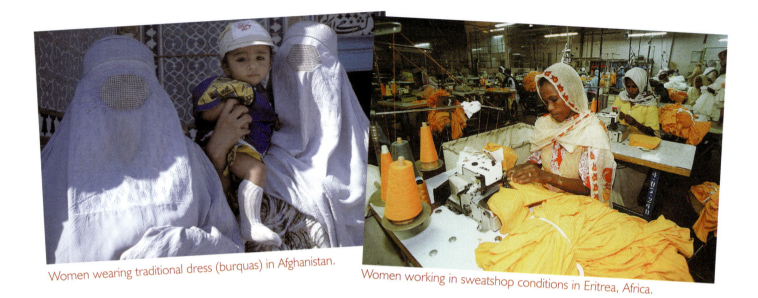

Women wearing traditional dress (burquas) in Afghanistan.

Women working in sweatshop conditions in Eritrea, Africa.

Class activity

In what ways do these images of women suggest discrimination against women in less-developed countries?

Slaves of the twenty-first century

An extreme form of human injustice occurs in the form of *slavery* of millions of men, women and children. Today, an estimated 27 million people in the world have been bought, sold, held captive and brutally exploited for profit.

In India, for example, millions of poor people have been caught in a *debt trap* leading to slavery. Unable to repay loans to moneylenders, they are forced to sell themselves in a form of slave labour. An estimated 15–20 million people are debt slaves in India, Pakistan, Bangladesh and Nepal.

For women, *prostitution* has long been a form of slavery. Today, criminal gangs illegally traffic in women. For example, more than 100,000 women have been moved illegally into Europe from eastern Europe, Africa and the Middle East. They end up mostly as prostitutes in Europe's major cities. And in India, some 50,000 prostitutes work in Mumbai, more than half trafficked illegally from Nepal. Their life expectancy is less than 40 years.

TEST YOURSELF AT

my-etest.com

SECTION 3 (CHAPTERS 8–13)
THE GLOBAL ECONOMY

By the start of the twenty-first century, a single, interdependent global economy had emerged. Central to this has been the growth of powerful multinational companies (MNCs) and their large-scale investments in different areas of the world. As a result, different areas have developed different roles within the global economy. This has given rise to a large growth in global trade and an international division of labour.

This section has six chapters:

- Chapter 8   Globalisation
- Chapter 9   Multinational Companies
- Chapter 10   Multinational Companies in the European Union and Ireland
- Chapter 11   Patterns of World Trade 1: Merchandise Trade
- Chapter 12   Patterns of World Trade 2: Services
- Chapter 13   The International Division of Labour

Rotterdam: Europe's largest port provides trade links to the rest of the world.

Even remote areas of the world are now affected by multinational companies.

CHAPTER 8
GLOBALISATION

KEY IDEA!

The world has become a smaller place in which decisions and events in one part of the world can have major impacts on people living in other parts of the world.

The world today appears to be changing more dramatically and more quickly than ever before. Furthermore, our lives seem to be influenced increasingly by events and decisions that occur far from where we live and work. The term **globalisation** is used to define such processes.

Globalisation is a process by which events, actions and decisions in one part of the world can have significant consequences for communities in distant parts of the globe. In effect, this means we are living in an increasingly and highly interconnected world.

The twin towers of the World Trade Center in New York were a symbol of US economic and political power. Their destruction in a terrorist attack on 11 September 2001 was to have major consequences for the world. Can you think of any of these consequences?

CAUSES OF GLOBALISATION

The process of globalisation, which allows people and places located throughout the world to interact more, is influenced by *five* main factors.

1. **Improvements in transportation** have reduced the costs and time involved in moving large amounts of raw materials, goods and people over longer distances e.g. bulk shipping, containerisation, air travel.

Supertankers of up to 500,000 tonnes play a major role in globalisation by transporting vast amounts of oil from oilfields to markets throughout the world. Why is this important for global development?

2. **Advances in telecommunications**, such as electronic mail (e-mail), communications satellites and fax machines link people and places more efficiently than older systems, such as the telephone and telegraph. Furthermore, the internationalisation of television, such as CNN, Sky and MTV, allows images of other places, events and cultures to be transmitted directly into our homes.

3. **The reorganisation of business** has seen the emergence of very large, international companies. These are termed multinational companies (MNCs), such as Ford, Microsoft, Sony, Nike. They purchase inputs, produce goods and services, and sell them at a global level.

> Look at your TV guide to see the origin of your favourite programmes. How many are produced in Ireland?

4. **Global banking and integrated financial markets**. Using new technologies to move capital through the world, about US$100 billion worth of currencies are traded daily. To control this trade, international banking and financial institutions have emerged, with headquarters in world cities such as London, New York, Frankfurt and Tokyo. Stock exchanges in these financial capitals have a key role to play in shaping development trends throughout the world.

5. **A more liberal world trading system** has been promoted strongly by richer capitalist countries. These have encouraged the movement to free trade by removing/reducing barriers that limit trade, such as tariffs and quota restrictions. This makes it easier for them to increase their dominant role in trading goods and services at a global level.

Financial dealings inside the London Stock Exchange involve traders moving vast amounts of money. This influences the economies of companies and countries throughout the world.

ECONOMIC GLOBALISATION

If we focus attention on economic globalisation, two aspects are critical for its definition and understanding. These aspects will also form the basis for the following five chapters.

● Multinational Companies and their Foreign Investment

A major cause, as well as a consequence, of economic globalisation is the large movements of capital (money) that occur at the global scale. One of the main forms of capital movements at this scale is **foreign direct investment** (FDI). This involves companies investing often large amounts of money to set up economic activities, such as mining of raw materials, factories or offices in a number of different countries.

Through improved transport and communication systems, they are able to co-ordinate their production and sales as part of a **single global system**. These companies with facilities in more than one country are termed **multinational companies (MNCs)** and are key elements of economic globalisation (see Chapter 9).

So, to recap, the two key components that define and drive the global economy are:
- multinational companies and their foreign direct investment
- the growing volume and value of international trade.

● Increased International Trade

High and increasing levels of international trade form the second key component of economic globalisation. It is a process, therefore, in which more and more goods and services produced in one country are sold in the markets of other countries (see Chapters 11 and 12).

A cartoonist's view of the process of globalisation. Study the cartoon: How many US products can you identify? What do you think is the main message illustrated by this cartoon?

How does this photograph of people in India collecting water from a communal tap illustrate the two key components of economic globalisation?

Class activity

How does the above photograph illustrate the two key components of economic globalisation?

TEST YOURSELF AT

my-etest.com

CHAPTER 9
MULTINATIONAL COMPANIES

KEY IDEA!

Multinational companies invest in foreign locations to reduce their costs of production and increase profits.

Since the 1960s, multinational companies (MNCs) have emerged as one of the most important elements shaping the global economy. As part of their strategy for growth, they undertake investments in countries which are outside their home country. This is termed **foreign direct investment** (FDI). Through this investment, MNCs are able to control increasing amounts of global trade, production and employment. They have, therefore, a major influence in the development prospects for most countries, including Ireland.

A **multinational company** is a company with production and service activities located in more than one country.

By 2000:
- some 44,000 MNCs operated globally
- these controlled about 380,000 subsidiary companies in foreign countries
- they accounted for almost one-third of the value of global exports.

THE GROWTH OF MNCS AND FDI

Foreign investment by MNCs began to show a major increase since the mid-1980s (Figure 9.2 on page 57). In 2000, their investments amounted to some $850 billion.

At the start of the twenty-first century, some of the largest MNCs had reached a scale of business which made them economically more powerful than many countries (Tables 9.1 and 9.2). So, the business of Exxon-Mobil, the MNC with the largest sales in 2001, approached an amount equal to the GNP of Belgium, and was almost one-half that of India, the second-most populous country in the world. Compared to the poverty-stricken countries of sub-Saharan Africa, these powerful MNCs had sales values that greatly exceeded their national GNPs.

The combined GNP of the 36 countries with the lowest values of the UN Human Development Index amounts to only $265 billion. Most of these countries are in sub-Saharan Africa.

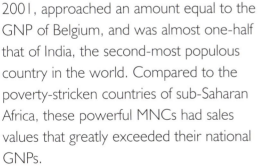

The headquarters of Daimler Chrysler in Stuttgart. This German multinational company took over Chrysler, the third-largest auto manufacturer in the USA in 2001. It is now the largest non-American MNC.

USA	Sales ($billion)	Employment ('000s)
1 Exxon-Mobil	206	180
2 Walmart	191	1,192
3 General Motors	185	392
4 Ford Motors	170	346
5 General Electric	130	312

Rest of the world	Sales ($billion)	Employment ('000s)
1 Daimler-Chrysler	150	450
2 Royal Dutch-Shell	149	95
3 BP	148	98
4 Mitsubishi	127	42
5 Toyota Motors	121	216

Table 9.1 Sales and employment levels in the five leading MNCs in the USA and rest of the world

GNP of selected countries (US$ billion)			
Germany	1873	Ireland	94
India	457	Czech Republic	51
Belgium	226	Ethiopia	7
Norway	162	Niger	2

Table 9.2 GNP of selected countries

Class activity

Study Tables 9.1 and 9.2.
1. Name the MNCs with (a) the largest sales; and (b) the highest employment.
2. What are the main products associated with each of the MNCs?
3. Why do you think these products are so dominant?
4. Identify the countries of origin of the MNCs listed under 'rest of the world'.
5. Does Ireland's GNP exceed the sales values of these MNCs?
6. Estimate how many times the sales value of the leading MNCs exceed the GNP of Ethiopia and Niger.

Reasons Why MNCs Locate Activities in Different Countries

MNCs benefit significantly from investing in foreign locations. This is linked to *four* main reasons:

● **Access to raw materials**

Historically, this was the most important reason. Colonial powers, such as Britain and France, invested heavily in their colonies to access raw materials, such as copper, coffee and rubber. MNCs continue to invest strongly in **securing supplies of critical raw materials**, such as oil and mineral ores. Many of these raw materials are sourced in less-developed countries, and MNCs use bulk shipping to transport them to processing plants often located in more-developed countries. Good examples include oil and iron ore shipped to refineries and steelworks located around the coasts of western Europe.

An oil drilling rig along the Nigerian coast. These rigs are usually owned by MNCs who ship most of the crude oil to refineries located in western Europe. Do you think these activities are of major benefit to less-developed countries? Why?

● Access to cheap labour

As mechanisation of production has reduced the importance of skilled labour, MNCs search for locations which offer **cheaper and more flexible sources of labour**. Many of the less-developed countries have large numbers of people who are prepared to work long hours for low wages. Movement of the production of basic or simple goods and services to less-developed countries generates considerable savings for MNCs through lower labour costs. This **relocation of production** has given rise to what is termed a **new international division of labour** (see Chapter 13).

Large numbers of female workers and also children are prepared to work for very low wages in less-developed countries. In what stage of the product cycle is this garment factory?

● Access to new and expanding markets

In today's world, three global markets dominate. These are North America (especially the USA), Japan and the European Union. To ensure **access** to these **prosperous markets**, most MNCs have a strategy of locating major production and servicing facilities in each of these regions.

The large Toyota car plant at Burneston in England. This opened in 1992 and manufactures some 300,000 vehicles a year, primarily for the EU market. Why do you think companies such as Toyota invest in countries like Britain?

Note the key role of labour for branch plant locations and the *product cycle* (see page 54).

● Flexibility of location

By having economic activities located in a number of different countries, MNCs are able to **move production between plants**, if it is deemed profitable to do so. So, for example, if labour costs rise too rapidly in Ireland and/or another country provides higher grants, then an MNC may decide to relocate part or all of its production from Ireland to the cheaper location. This gives MNCs considerable bargaining power with national governments.

The Product Cycle and Changing Locations of MNCs

The reasons which influence MNCs to invest in foreign locations can be viewed as part of the **product cycle**. This suggests that products evolve through **four** stages, and that each stage is influenced by a different set of location factors. So, as a product evolves from its first stage, when it is being researched and developed, to the final stage when it is mass produced as a basic good, the preferred locations change (Table 9.3).

The product cycle		
Stage	*Location factors*	*Preferred location*
1. Development of new product	Access to scientists, technology, investment money	Core urban regions, e.g. London, Boston, Tokyo
2. Early growth	Skilled workers, good regional market and infrastructure, e.g. business services, finance, marketing	Growth regions such as the Manchester–Milan axis, Catalonia
3. Rapid growth/ maturity	Less-skilled and lower-cost labour, cheaper land and good transport systems to access world market	Peripheral areas of more-developed countries or NICs, e.g. Ireland, South Korea, India
4. Old age or stagnation	Low labour costs, flexible workers, cheap land and government grants, access to world market	Less-developed countries in peripheral world regions such as south and east Asia

Table 9.3 The product cycle

Class activity

Study Table 9.3.
1. Explain the different location choice for each of the four stages of the product cycle.
2. In what stages would you expect to find branch plants? Why?
3. Why do you think MNCs adopt a strategy of opening/closing branch plants? (Hint: competition and low costs).
4. In the 1960s, Ireland was in Stage 3 but also attracted Stage 4 products. Now it is in Stage 2, although the Dublin region especially has elements of Stage 1. Can you explain this evolution of Ireland through the product cycle?

In the early stages, the main needs are for highly skilled workers and technology to develop the new product. These occur mainly in the major urban centres of developed regions, which also provide prosperous markets for the output. As the product becomes more basic in design, so competition increases and the concern of MNCs is to reduce costs of production. As a result, they move production from high-cost core regions to *branch plants* which are located in lower-cost peripheries.

By relocating production at different stages of the product cycle, MNCs benefit from the locational advantages of both core and peripheral regions.

Branch plants are basic production units of MNCs. They have little decision-making capacity and focus on the mass production of standard goods using relatively low-cost and low-skilled labour (see page 58).

THE GLOBAL ASSEMBLY LINE

The product cycle can also be seen to be a part of the global assembly line. This is a term used to describe the **global production and sales systems** of MNCs. In effect, **MNCs locate different activities in different places to achieve lowest costs/highest profits** (Figure 9.1). Generally:

● decision-making and the more skill-demanding and higher-value elements of the production system are located in the more highly developed countries, while

● basic production and services are relocated to less-developed countries where costs are lower.

> The production of a book, such as this, is an example of the global assembly line. The author in Cork writes the text, while editorial and graphic design work are located in London. Basic typesetting and production of the book occurs in south-east Asia. Books are shipped in bulk to the headquarters and sales offices of the publisher located in Dublin for distribution within Ireland and other parts of Europe.

Intel, the American electronics MNC, has invested heavily in its plant in Leixlip. In addition to producing a range of high quality products, it also includes research and development (R&D) and other critical services. It is Intel's major European facility. Would you describe this as a branch plant? In what stage of the product cycle would you position this plant?

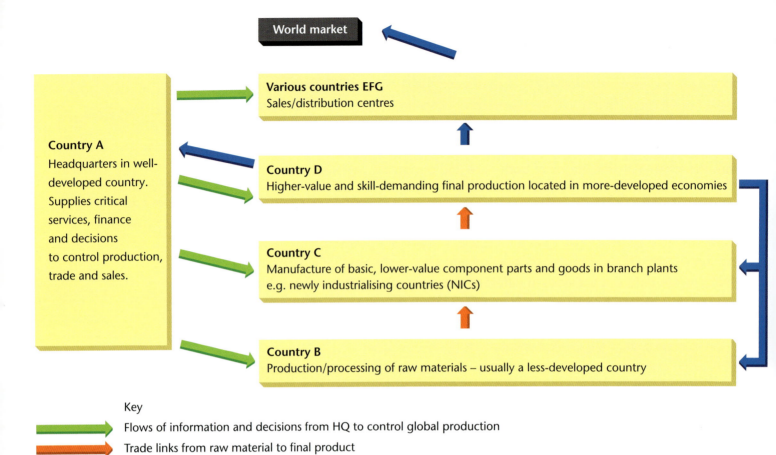

Fig. 9.1 Simplified model of the global assembly line of an MNC

Key

Flows of information and decisions from HQ to control global production

Trade links from raw material to final product

Trade in finished product

Class activity

Study Figure 9.1.

1. Why are the headquarters of most MNCs located in a well-developed country?
   Provide an example.
2. What advantages do newly industrialising countries (NICs) have for MNCs?
   (Refer to the product cycle, see pages 54–55.)
3. Explain why transport and communication links are so important for the global assembly line of MNCs.

Note that the global assembly line confirms the ideas introduced in Chapters 6 and 7 (especially that of fair trade) which favour development in core regions.

THE INVESTMENT AND LOCATION OF MNCS

There are *two* main types of location for MNCs

1. major industrialised regions
2. peripheral regions.

1. Major industrialised regions

Although MNC investment in less-developed countries has been increasing, most remains concentrated in industrialised countries, such as the USA, Japan and western Europe (Figure 9.2). This shows the powerful attraction of their prosperous markets for goods and services. By 2000, some 70 per cent of total MNC investments were located in the industrialised countries that contain only 15 per cent of the global population.

> Large and prosperous markets are the key attraction for MNC investment in the industrialised countries of the world.

2. Peripheral regions

Investment in developing countries has historically been low and was focused mainly on the exploitation of raw materials. Since the late-1980s, however, investment in such countries has increased significantly (Figure 9.2). By 2000, some 25 per cent of the world's manufacturing production was located in developing countries. This compares with less than 10 per cent until the late 1960s (Figure 9.3).

This is due to the dispersal of **branch plant activities** to cheaper locations in the global periphery. One important result has been to help industrialise a growing number of developing countries. Where this has been particularly successful, such as in Brazil, India, South Korea and Thailand, the term **newly industrialising country** (NIC) is used.

> The main locational advantages in attracting MNCs to less-developed economies are the low costs and the flexible attitude of the workforce.

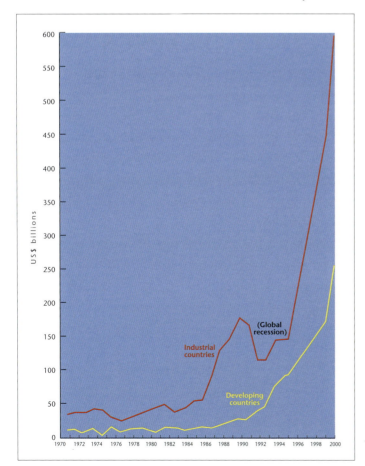

Class activity
Study Figure 9.2.
1. Estimate the total amount of MNC investments in 1970 and 2000.
2. Which group of countries receives most investment? Why?
3. Why should the global recession cause a major decline in MNC investment for developed countries in the early 1990s?
4. Describe the trends for MNC investments in developing countries and industrial countries.
5. Suggest some reasons for the rapid increase of MNC investment to developing countries (refer to branch plants and new international division of labour).

Fig. 9.2 MNC investment flows to industrial and developing countries, 1970–2000

57

Some countries and world regions, however, continue to attract little MNC investment and do not show much improvement economically. This is seen clearly in sub-Saharan Africa, which remains the least industrialised global region (Figure 1.4, page 8).

> Why do you think sub-Saharan Africa attracts little multinational investment? Refer to page 18 to help you with your answers.

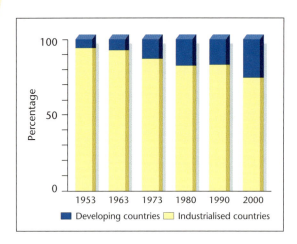

Fig. 9.3 The changing distribution of world manufacturing production between industrialised and developing countries, 1953–2000

Class activity

Study Figure 9.3.

1. Have developing countries increased their share of world manufacturing production since the Second World War?
2. What decade showed the greatest increase for developing countries?
3. Briefly explain what factors encouraged the growth of manufacturing production in developing countries.

> A **branch plant economy** is one with a high dependency on branch plants for employment and wealth creation. These economies tend to be unstable and show limited potential for longer-term growth and development.

BRANCH PLANTS OF MNCS

One of the main strategies of MNCs to remain competitive is to establish branch plants in peripheral regions of the world.

Branch plants can bring a variety of benefits to countries (Table 9.4). As a result, many governments, including that of Ireland, offer significant incentives, such as cash grants and tax benefits, to encourage MNCs to locate branch plants in their country.

Branch plants can also bring problems to host countries. This is particularly the case if **dependency** upon branch plants for employment and prosperity is too high. When this occurs, we use the term **branch plant economy**.

Probably the most significant problem is linked to the loss of control over decision-making. Decisions are often made at the headquarters of a MNC with little or no consultation with the workforce and government of the host country. Branch plants can, therefore, be closed or rundown if the MNC considers other locations to be a more profitable option. Overdependency upon branch plants is not regarded as an effective basis for sustainable development.

> So, what is considered to be the most important single problem linked with MNCs in countries such as Ireland?

Table 9.4 Advantages and disadvantages of branch plants

Advantages	Disadvantages
1. They provide many jobs.	1. Wage rates are often not high (late stages of product cycle).
2. They introduce new skills to workforce and new technologies.	2. Focus on basic production requires relatively low skill levels – repetitive and often boring work.
3. They can help increase demand for other goods and services located in the country (the multiplier effect).	3. Linkages with national suppliers of goods/services can be poor. Many key goods and services are purchased from other units of the MNC and in other countries (poor multiplier effect).
4. They bring a lot of investment and stimulate exports.	4. A lot of the profits are lost through repatriation to headquarters and shareholders outside of the country.
5. They diversify the economy and help integration into the global economy.	5. Instability – the type of work and the possession of branch plants by the MNC in other countries can cause a plant to be run down/closed and production transferred to cheaper and more profitable locations.

The competitive markets for clothing products of the 1990s caused Fruit of the Loom to close or run down most of its branch plants in Donegal. Most production was relocated to Morocco where labour costs were up to 90 per cent cheaper.

The ripple effect of globalisation: branch plant closure

Tradingworld Limited is an MNC headquartered in New York and opens a branch plant at Anytown in the west of Ireland. Initially, the company is successful and rapidly expands production and employment at its Irish branch plant.

This increases opportunities for small businesses in Anytown, while also attracting a significant number of business visitors to the plant. The population of the town grows rapidly, as workers from the surrounding area choose to build homes in Anytown to reduce the need for long-distance journeys to work. With a growing young population, the local GAA club extends its facilities and number of teams, and wins the county championship.

Fifteen years after opening its branch plant, recession and problems of rising costs cause Tradingworld Limited to rationalise its production. As part of its rationalisation, the headquarters in New York decide to close the operation at Anytown and relocate its entire production to a lower-cost factory in south Asia.

This causes the loss of 750 jobs directly employed at the Anytown plant, and some €50 million in wages, salaries and purchases which are spent mostly in and around Anytown.

Class activity

Read the fictitious account shown above of a branch plant investment in Ireland and discuss how you think the closure would affect the following activities/people in Anytown:

1. Tobin & Son, the local haulage contractor that had extended its business to serve the transport needs of Tradingworld.
2. Irishservices Limited, which had relocated its small business to Anytown to be close to Tradingworld, their principal market.
3. O'Flannagan's, the town's main hotel, together with other bars and restaurants.
4. Families who had taken out large mortgages to buy and build new homes in the town.
5. The local GAA club.
6. So, is overdependency on branch plants generally a good policy for development?

TEST YOURSELF AT
my-etest.com

CHAPTER 10
MULTINATIONAL COMPANIES IN THE EUROPEAN UNION AND IRELAND

KEY IDEA! Large-scale investments by MNCs in both the EU and Ireland have given rise to complex trading patterns and helped increase their economic development.

THE EUROPEAN UNION

The European Union (EU) forms one of the richest regions of the global economy. As a result, many MNCs have been attracted to the EU. In addition, however, MNCs headquartered in the EU have also invested heavily in other parts of the world. These trends have been a major factor in promoting trade and the development of this prosperous global region.

EUROPEAN MNCS AND GLOBAL INVESTMENT

Multinational companies from the EU, and especially its core countries in north-west Europe, have a long tradition of foreign investment. This role increased strongly in the 1990s, when the EU replaced the USA as the main source region for MNC investment. Most of this investment is directed to other prosperous and developed world regions. An increasing amount, however, is used to open branch plants and service activities, such as back offices, in NICs in south-east Asia and Latin America. Former colonies of EU countries also attract investments.

Almost 50 per cent of MNC investment from the EU is directed to the USA.

Volkswagen's major car assembly plant in Germany. Comment on the importance of the railway line, sports stadium and apartment blocks located adjacent to the building.

Case Study: Volkswagen

Volkswagen is a major German MNC that has invested heavily in a number of foreign locations. As a result, this MNC has established vehicle assembly plants and component suppliers both in developed and less-developed countries.

Figure 10.1 shows the location pattern of VW's major investments in vehicle manufacturing. It also illustrates the complex trade flows that occur between its various locations and gives rise to the concept of a *global assembly line*.

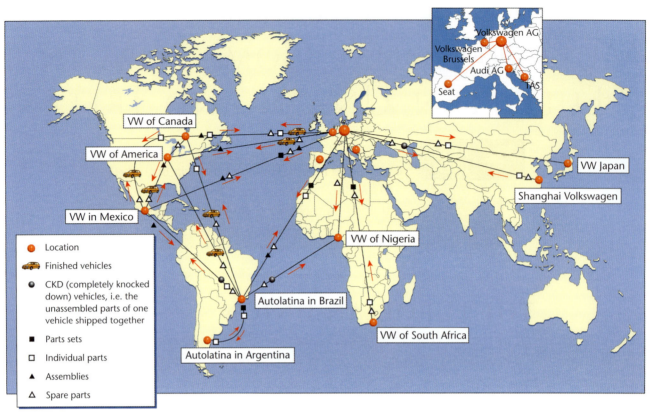

Fig. 10.1 Volkswagen's global assembly line

Class activity

Study Figure 10.1.

1. In what country are the headquarters of VW located?
2. In how many countries do VW have production plants?
3. Describe the trade flows that link the VW plant in Brazil with its other locations.
4. Why do you think that VW chooses to export cars to the North American market from Brazil and Mexico rather than from Europe?

MNC Investment in the EU

The European Union also attracts large amounts of investment from MNCs with headquarters outside the EU. This is due mainly to its large, prosperous and growing market. Almost 90 per cent of this comes from US and Japanese MNCs. Britain receives the highest national share of this investment, although peripheral countries such as Ireland and Spain also benefit significantly.

Ireland, with only 1 per cent of the population of the EU, receive some 8 per cent of MNC investment in the region.

Case Study: Ford Europe

Ford is one of the largest MNCs operating within the global economy. It organises its worldwide production system in terms of global regions. So, **Ford Europe** covers the vital European market and integrates its manufacturing and sales at this level.

Name of company:	Ford Motor
Headquarters:	Michigan, USA
Total world sales:	$170 billion
Global employment:	346,000

The production of vehicles is complex and involves the assemblage of a great variety of component parts and materials. Rather than producing most of these inputs themselves, Ford Europe subcontracts the work to specialist suppliers throughout Europe (Figure 10.2). In this way, Ford Europe is able to reduce its costs. This requires good transport links between the Ford vehicle assembly plants and suppliers. The suppliers also have to be able to guarantee the quality of supplies and deliver on time.

The most specialised and higher-value components are generally sourced in core regions, since they have the skilled workers and technology necessary to produce these inputs. In contrast, lower-value components can be located in peripheral regions where labour costs are lower.

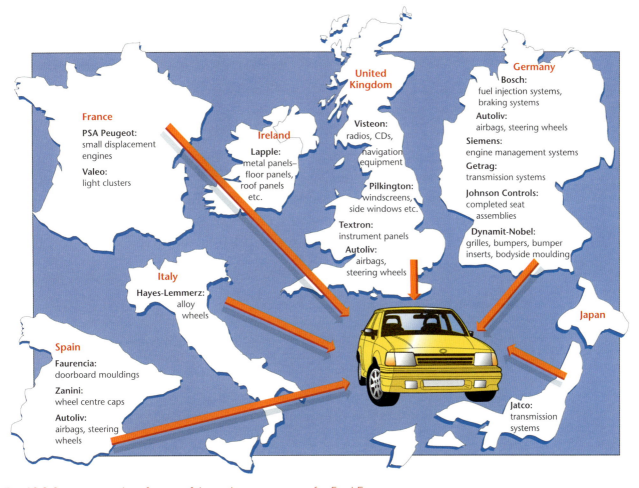

Fig. 10.2 Source countries of some of the major components for Ford Europe

Fig. 10.3 Location of main vehicle assembly plants of Ford Europe

In addition, Ford Europe's strategy for vehicle assembly is for its major plants to concentrate on one or only a limited range of its models (Figure 10.3). This allows each plant to benefit from lower costs associated with specialisation of production. Components are delivered to each plant from its specialist European suppliers. The finished cars are distributed through Ford sales offices located in all European countries.

Class activity

Study Figures 10.2 and 10.3.

1. In how many countries does Ford Europe have a vehicle assembly plant?
2. How many countries in Europe provide major supplies of component parts for Ford Europe?
3. Which country provides most component supplies? Suggest reasons why.
4. Suggest locational advantages of peripheral countries such as Spain and Portugal for vehicle assembly.
5. Why are good transport systems so important for MNCs such as Ford Europe?

Ford's major car production plant at Valencia in Spain. What would attract Ford to this location rather than developing its plants in the core of Europe?

Since Ford closed its car assembly plant at Cork in 1984, all its cars sold in Ireland are imported through the port of Cork from its plants elsewhere in Europe.

Multinational Companies in Ireland

Of all EU countries, Ireland has benefited most from the inward investment of MNCs as a basis for economic development. From the 1960s, the government focused on attracting MNCs to increase national wealth and employment. This policy of *industrialisation by invitation* has been successful.

Refer to pages 21–24 and to *Our Dynamic World 1*, Chapter 18, for further information on industrial development in Ireland.

The Pfizer chemical plant at Ringaskiddy in Cork Harbour was one of the first of a new wave of MNCs attracted to Ireland in the 1960s. Why did such MNCs move to Ireland?

65

Branch plants 1960s–1980s	High-tech plants 1990s –
● Plenty of cheap but unskilled labour	● Large supply of well-educated and young workers
● Cheap land (rural areas)	● Access to good-quality research institutions and universities
● Government grants	● Low taxation rates for industry
● Access to EU market	● Access to expanding EU market
● Basic services and adequate transport systems	● High-quality environment, e.g. housing, recreation

Table 10.1 Factors attracting branch plants and high-tech MNC investment to Ireland

From the 1960s to the 1980s, large numbers of *branch plants* were attracted to Ireland (Table 10.1). Ireland became a *branch plant economy*.

Recession and increasing competition for branch plants in the 1980s, however, resulted in MNCs closing factories/reducing employment in Ireland, and looking for even lower cost locations in the less-developed world. This highlighted the problem of *dependency* on branch plants, and on decision-making at headquarters of MNCs which are located outside of Ireland.

Why should dependency on branch plants be a problem for long-term national development? Review Table 9.4 on page 59.

Class activity

Use Table 10.1 to explain why:
1. More recent MNC investment locates mainly in larger urban centres, especially Dublin.
2. Branch plants can be closed and production relocated to less-developed countries.
3. Higher-tech plants are less likely to be closed/relocated to less-developed countries.

Do these factors suggest newer MNC plants are at a different stage of the product cycle than more traditional branch plants? Explain.

In the 1990s, a *new wave of MNC investment* was attracted to Ireland. Many of these differed from earlier branch plant operations and were a vital factor in developing the Celtic Tiger economy.

● Most investment is concentrated in *growth sectors*, such as electronics, pharmaceuticals and internationally-traded services.

● The focus is on producing *high-value goods and services* rather than mass production of low-value goods in branch plants.

● New plants include *more key decision-making functions*, such as research and development, marketing and finance.

● MNCs are *less likely to close* their newer plants than branch plants. Why?

Today, more than one in every two industrial jobs in Ireland are provided by foreign-owned MNCs. The Dublin urban region, in particular, has benefited most from this new wave of high-tech investment.

In 2001, US investment in Ireland amounted to $33 billion with some 600 American firms employing 95,000 workers.

The University of Limerick is the main reason for the location and success of the adjacent Plassey Technology Park. What benefits are there in locating businesses near universities?

Case Study: Dell Computer Corporation, Ireland

Name:	Dell Computer Corporation
Headquarters:	Austin, Texas
Main product:	Computer systems
Global employment:	39,000
World sales:	US$35.5 billion

Dell is one of the world's major manufacturers of computer systems. With its headquarters in the USA, Dell has established a global system of production and sales. To help organise this system, Dell has divided the world into three global regions. Each has its own regional headquarters, production and sales network (Figure 10.4).

Dell Ireland

Dell opened its first Irish operation in Limerick in 1991 with only 120 employees. Since that date, the company has expanded rapidly in both Limerick and the Dublin area.

Limerick is the production centre for Europe, the Middle East and Africa (EMEA). Its two plants employ some 3,300 people, and the company is the most important employer in the Limerick region.

Dell's main production plant at Raheen, Limerick. The company has opened a second production facility in Limerick at the Plassey Technology Park adjacent to the University of Limerick.

Limerick is the largest production facility for Dell outside of the USA.

In the **Dublin** region, Dell has a further two plants at Bray (1992) and Cherrywood in Cabinteely (2000). These act as sales centres for Dell Ireland. In addition, Bray is a major call centre for Dell, dealing with sales queries and service problems from its international market. Cherrywood is the administration centre for Dell in Ireland. Together, these two plants employ some 1,200 people.

Reasons for Dell locating in Ireland

- High-quality education systems and access to well-educated and skilled workers. The University of Limerick is especially significant for this high-tech industry.
- Availability of suitable sites such as the industrial estates at Raheen and the Technology Park at Plassey in Limerick.
- Well-developed telecommunications systems. This is vital for the decision to locate the call centre at Bray.
- Good infrastructure to export finished products and import component parts e.g. Shannon Airport, road access to Dublin and Rosslare.
- Strong government support – grants and low corporation tax.
- Proximity to other world-class computer companies such as Intel, Microsoft, 3com.
- English-speaking country.

Sources of raw materials and components

The vast majority of inputs required by Dell Ireland have to be imported (Figure 10.5). Only 10 per cent are sourced in Ireland. This highlights the global assembly line of Dell.

Markets

Limerick is Dell's production centre for the extensive global region covering the EMEA. The company has a large number of strategic sales offices throughout this region (Figure 10.4). Countries in western Europe, such as Britain, France and Germany, are the main markets for Dell.

Impact of Dell in Ireland

- 4,500 employed directly
- approximately 30,000 indirect jobs
- 5.5 per cent of total value of Ireland's exports
- 2 per cent of the country's GDP
- high profile for country in high-tech industry
- significant expenditure and capital investments.

Class activity

Approach an MNC located in your region and find out:

1. Why they chose their location.
2. The global pattern of production plants.
3. Their principal sources of raw materials and component parts.
4. The location of their main markets.
5. Their main impacts on your region.

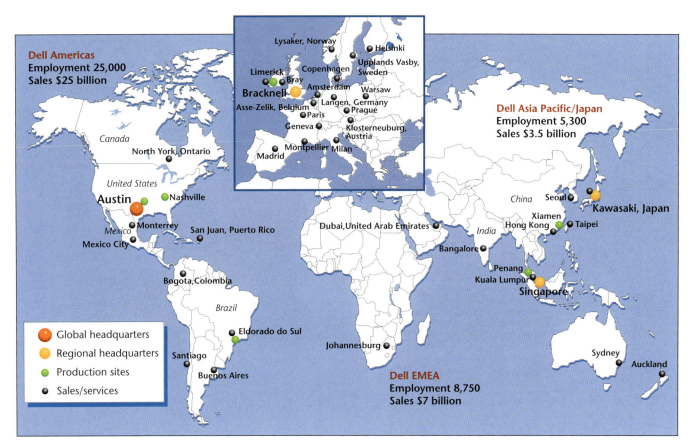

Fig. 10.4 Dell's global production system

Class activity

Study Figure 10.4.

1. Where is the global headquarters of Dell?
2. How many production sites does Dell operate?
3. Where is the location of Dell's only production site for its EMEA global region?
4. How many sales offices does Dell have in its EMEA global region? Can you suggest any reason as to why there are so many?
5. Why do you think the Middle East and Africa are each covered by only one sales office?
6. Suggest reasons why the regional headquarters for the EMEA region is located near London rather than in Ireland.

So, why is it important to attract MNCs like Dell to Ireland?

Fig. 10.5 Sources of raw materials and components for Dell (Limerick)

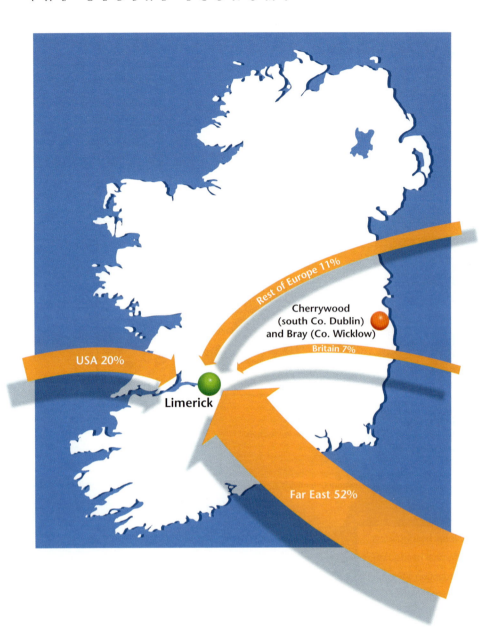

Rest of Europe 11%

Cherrywood
(south Co. Dublin)
and Bray (Co. Wicklow)

USA 20%

Britain 7%

Limerick

Far East 52%

Class activity

Study Figure 10.5.

1. What world region supplies most inputs?
2. Can you suggest reasons for this? (Hint: NICs, branch plants and NIDL, see Chapter 13.)
3. What percentage of inputs are obtained within Ireland?
4. Why is it important to increase this amount?
5. Suggest reasons why improvements to Ireland's transport and communications systems have been essential for the success of MNCs such as Dell.

TEST YOURSELF AT
my-etest.com

CHAPTER 11
PATTERNS OF WORLD TRADE 1: MERCHANDISE TRADE

KEY IDEA!

World trade in merchandise goods has increased strongly since the Second World War, although much of this trade is focused on three dominant global regions: **the global triad.**

Since the end of the Second World War, the volume and value of world trade has increased dramatically. **Two** key factors have encouraged this growth and have emphasised linkages between different regions of the world economy.

1. **The growing number and power of MNCs** which organise production at a global level. This has led to an increasing trade in goods/raw materials, and flows of information/services, between different countries/regions of the world.
2. **Innovations in transportation and communications** have been vital to accommodate the growth in global trade flows.

Remember the five factors that influence globalisation. See pages 48–49.

Large sea ports, such as Rotterdam, have made the EU the most important world region for merchandise trade. Why are sea ports so important for world trade in merchandise goods?

71

MERCHANDISE TRADE

Merchandise trade has grown rapidly as different countries specialise in producing goods for which they have a comparative advantage.

Not all regions have shared equally in the increases in world trade (Figure 11.1 and Table 11.1). Some world regions, such as sub-Saharan Africa and South Asia, remain marginal to this process. This is because of:

- their **late development**, linked to their colonial history
- their concentration on the **products with relatively low value** e.g. raw materials and/or mass-produced and basic goods manufactured in branch plants of MNCs
- the prices of most primary commodities, which have declined compared to prices of higher value goods produced in developed world regions. This is referred to as the **terms of trade** (see page 40).

> Merchandise trade involves trade in goods that have a physical form, such as food products, raw materials and manufactured goods.

> Review what is meant by **fair trade** (see pages 40–41).

Why do most poverty-stricken countries of sub-Saharan Africa have little role in world merchandise trade?

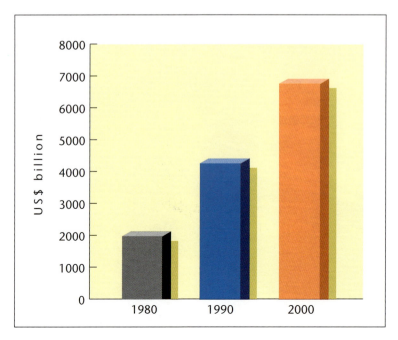

Fig. 11.1 Value of world merchandise trade, 1980–2000

Region	1990	2000
East Asia and Pacific	240	614
East Europe and central Asia	189	381
Latin America and the Caribbean	169	336
Middle East and North Africa	134	132
South Asia	34	71
Sub-Saharan Africa	80	90
Low-/middle-income countries	834	1,548
High-income countries	3,418	5,183
World total	4,252	6,767

Table 11.1 Merchandise exports by global region, 1990–2000 (US$ billion)

Class activity

Study Figure 11.1 and Table 11.1.
1. Describe the trend in world trade from 1980 to 2000.
2. What conclusions do you make from this trend in merchandise exports?
3. What two regions show the lowest involvement in merchandise exports? Suggest some reasons for this.
4. Identify the three world regions that more than doubled their merchandise exports in the 1990s. Suggest some reasons for this.

Perhaps the most outstanding feature of world trade is the dominance of high-income regions. In effect, **three core economies** organise and control trade in merchandise goods. These are the EU, the USA and Japan. Together, these are termed the **global trading triad** (Figure 11.2).

The global trading triad controls well over half of the value of world merchandise trade. The EU is the most dominant trading partner within the triad, while trade between its member states exceeds that with the USA and Japan.

Some 75 per cent of merchandise trade is sourced in high-income regions. These regions contain only 15 per cent of the world's population.

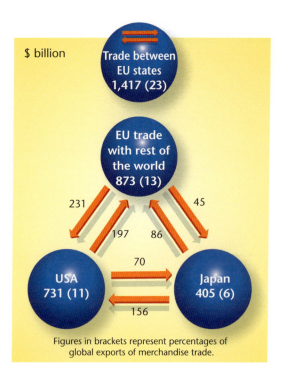

Fig. 11.2 The global trading triad (US$ billion)

$ billion

Trade between
EU states
1,417 (23)

EU trade
with rest of
the world
873 (13)

231   45

197   86

70

USA
731 (11)

Japan
405 (6)

156

Figures in brackets represent percentages of
global exports of merchandise trade.

Class activity
Study Figure 11.2.
1. What percentage of world exports is sourced in these three regions?
2. Which core region is the largest exporter?
3. Does the EU trade more between its member states than with the USA and Japan?
4. Between which two core regions does the greatest trade occur? Suggest reasons for this.

Balance of trade means the difference in the value of a country's imports and exports. Why is a positive balance of trade important for development?

Japan's role in world trade declined significantly in the 1990s due to a major economic recession in that country. In spite of this, it still achieves a strong positive **balance of trade** within the trading triad. This contrasts strongly with the position of the USA.

Japan's prosperity since the Second World War has been based mainly on its merchandise trade. What evidence is there in this photograph to suggest that Japan is a major exporter of merchandise goods?

The global trading triad is, in effect, 'sucking in' more and more of the world's productive activity, trade and investment. The triad sits astride the global economy like a modern three-legged colossus. It constitutes the world's mega markets.

Class activity

From your study of MNCs and merchandise trade flows, do you think the above statement is an accurate description of world economic development? Explain.

TEST YOURSELF AT
my-etest.com

CHAPTER 12
PATTERNS OF WORLD TRADE 2: SERVICES

 **KEY IDEA!**

Globalisation has increased demands for a wide range of services, and international trade in services has become a vital element for an efficient global economy.

Quaternary services involve the collection, processing and transmission of information and knowledge and therefore generally require a well-educated workforce. Examples include legal and financial services, marketing and research and development.

The Internet has helped to make the world a 'smaller place.' Explain this statement.

Almost 80 per cent of Internet users live in the USA and Europe.

The globalisation of industrial production and the growing importance of MNCs have increased the demands for a wide range of services. Demand is especially high for **quaternary services**. These are recognised as the global growth sector and are responsible for an increasing amount of employment and wealth creation. As a result, the services sector and trade in services has become an increasingly important element in world development and trade.

Innovations in communication systems have been vital for the large-scale and rapid developments in global services. The telephone, fax, satellite communication links and, above all, the Internet, allow large volumes of information to be transmitted around the world. At the same time, efficient computer systems are able to store and process such data. These innovations have been developed and are used most in core world regions. This has allowed such regions to further extend their control over the world peripheral regions (see Figure 12.3 on page 80).

Do you have e-mail? Why are Internet cafés becoming increasingly popular?

As the importance of services has increased throughout the world, service industries have internationalised their activities. So, in a similar way to branch plants, a growing number of service companies are relocating some of their **back-office functions** to peripheral regions. Providing communication systems are adequate, this allows them to take advantage of lower labour costs.

The decentralisation of services has tended to benefit only those countries which have invested in upgrading their telecommunications and education systems. This is vital for the successful take-off of this sector. Examples include India and Ireland.

For most countries in the global periphery, the costs and levels of education needed to support office-related functions remain an effective barrier to their development. As a result, **the relocation of international service functions to the developing countries is much less extensive than it is for branch plants of manufacturing industries**.

An employee at a call centre in Delhi, India, answering questions from customers in the USA by e-mail. What advantages does India have for call centres?

OFFSHORE FINANCIAL CENTRES

As part of globalisation, vast amounts of money circulate through the world economy. While traditional banking centres deal with the majority of these flows, they are less able to meet the needs of clients who look for secrecy and/or shelter from taxation and other forms of regulation. The result has been the growth of **offshore financial centres**.

These are generally islands or microstates, such as the Isle of Man and Andorra in Europe, which have become specialised centres for trade in international finance. Vast amounts of capital flow through these centres and create significant wealth and employment. One concentration of such centres occurs in and around the Caribbean. Examples include the Cayman Islands (population, 36,000), which have some 600 banks with deposits estimated at over $450 billion. Furthermore, in 2000, over $900 billion passed through the financial institutions in the Caymans, making this small group of islands the most successful of all offshore centres.

A back office deals with the provision of a range of services which generally involves the use of many workers who process large volumes of basic information. Examples include processing of insurance claims, hotel and travel reservations.

The Cayman Islands are a major centre for offshore banking. Use the photograph to suggest another important international industry that occurs in these islands.

New York with its powerful stock exchange, large number of business headquarters and excellent communications networks is one of the world's major geographical centres of control. What does this mean?

Geographical centres of control are major world cities that dominate international flows of finance, information and decision-making. The headquarters of many MNCs are located within them.

Although some relocation of services to the global periphery has occurred, the dominant characteristic of this sector is its continued focus on core regions. And, within these core regions, it tends to be the largest international cities that have attracted a growing share of the services sector, especially higher-value services. These cities are called **geographical centres of control**, and include London, New York, Tokyo (Figure 12.1).

The continued dominance of the global core regions is also highlighted by the communication flows that occur between them (Figure 12.2).

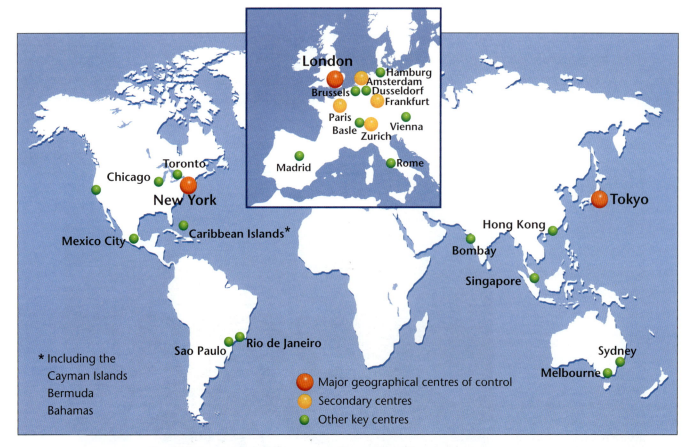

Fig. 12.1 Locations of major centres of international finance

Fig. 12.2 Communication flows between major world regions

North America (26.7)

4.7   4.5

Asia and Oceania (including Japan) (15.1)

1.6   2.5

2.7   1.5

0.02   0.3

Europe (35.2)

Latin America and the Caribbean (3.3)

0.1   0.1

0.7

1.0

Above shows total annual traffic from one region to another (in billions of minutes)
Arrows indicate direction of traffic between regions
Numbers in brackets indicate the total amount of international traffic for countries within that global region

Class activity

Study Figures 12.1 and 12.2.

1. Name the three major geographical centres of control for international finance.
2. In which global region are located the most international financial centres? Suggest some reasons for this.
3. Why are there no significant financial service centres in Africa and much of Asia?
4. Which core region dominates flows in telecommunications? Suggest reasons for this (Hints: prosperity and location).
5. Does Figure 12.2 support the concept of a global trading triad (refer also to Figure 11.2 and 12.3).

A street scene in Mali, sub-Saharan Africa. What evidence is shown to suggest that demand for international services is not high in such underdeveloped economies?

As the world settles into the twenty-first century, the development gap between the world's core and peripheral regions remains large. At the heart of these 'two worlds' are the three core regions which form the **global trading triad**. These core economies have become more strongly linked through trade, investment and improvements in transport and communication systems. In addition, they have used trade and investment to increase their influence over poorer regions of the world.

Figure 12.3 illustrates the three core regions of the global trading triad and their preferred areas of extended influence.

Fig. 12.3 The three core regions of the global triad and their areas of preferred global influence

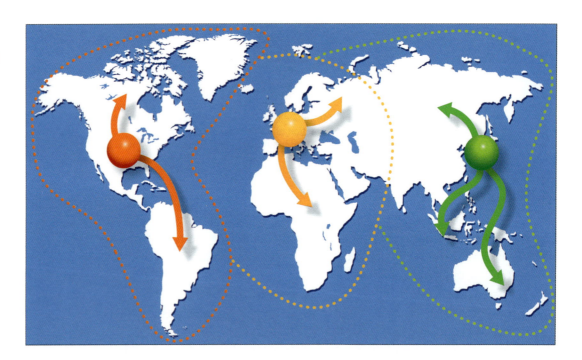

Class activity

Study Figure 12.3.

Select a core region and discuss the reasons why it has been able to extend its influence over certain regions of the world periphery. Is this likely to benefit the core or peripheral world regions?

- **The USA** is extending its influences through North and South America. The North American Free Trade Area (NAFTA) is an attempt to create an organisation to link these global regions more effectively.
- **The European Union** has extended its influence into central and eastern Europe through enlargement (see *Our Dynamic World 1*, Chapter 27). It is also strengthening ties with Russia, while its colonial links with Africa and the Middle East give Europe a strong presence in these regions.
- **Japan** is extending its influence through much of south-east Asia and into Australia/ New Zealand. The large market of China and resources of eastern Russia are also of interest to Japan.

TEST YOURSELF AT
my-etest.com

CHAPTER 13
THE INTERNATIONAL DIVISION OF LABOUR

KEY IDEA!

The workforces of different countries tend to specialise in producing particular goods and/or services. This international division of labour is an essential part of globalisation.

All economic activities require labour. Furthermore, development theory suggests that if workers are allowed to specialise in certain tasks or products, then they can become more efficient. Productivity rises, while the costs of production fall. This concept is known as **the division of labour**.

Closely linked to this concept is that of **the spatial division of labour**. This argues that further economic benefits can be gained if specialisation of production occurs in different areas. It allows the workforce in such areas to concentrate on the production of certain goods and services. In addition, land use and infrastructures become closely linked with the specialised production. This raises productivity further while lowering costs.

So, for example, in a city, the CBD (central business district) provides specialist services, such as banking, legal and insurance, the retail area provides major shopping facilities, while manufacturing activities are often located on industrial estates adjacent to ring roads around the edge of the city.

In other words, the more familiar a worker becomes with a task, the easier it is for the worker to perform that task. So, productivity rises while costs fall.

Can you identify specialist areas of employment in your home area?

Henry Street in Dublin. What is the main employment provided in areas such as this?

At the national level, the spatial division of labour provides many benefits. Economic development and productivity increases as a country focuses on producing goods and/or services for which it has a **comparative advantage** over other areas. This allows that country to export these goods/services. The profits made from export trade allow it to import those goods/services which the country does not produce, or can produce only at costs which are higher than in other countries. Specialisation of labour at this level is called the **international division of labour**.

What is the comparative advantage of countries such as Saudi Arabia? How does this help them promote economic development?

Comparative advantage means that a region possesses some advantage, such as raw materials or a specialised labour force, which allows it to produce particular goods more cheaply than other regions. This is vital for success in world trade.

THE EVOLVING INTERNATIONAL DIVISION OF LABOUR

The idea of an evolving international division of labour is central to the emergence of the global economy. In particular, it is linked to the **two key features** which have driven the process of globalisation:

- MNCs and their investments in branch plants and services.
- increased levels of global trade.

Since the start of the industrial revolution, the international division of labour has gone through three phases, with a fourth phase just beginning (Figure 13.1):

Fig. 13.1 Evolution of the international division of labour

	Dominant employment	Core	Trade	Periphery	Dominant employment
Phase 1 (1800–1960s)	Large-scale industrial development		industrial goods → ← raw materials, food		Primary activities such as mining and agriculture
Phase 2* (1970s–80s)	Higher-value industries and services		high value goods and services → ← raw materials and low-value, basic industrial goods		primary activities but growing branch plant activities
Phase 3* (1990s–)	High-tech industries and quaternary services		high-tech goods and services → ← basic industrial goods and standard services		branch plants and back-office services
Phase 4* (2000–)	High-tech industries and quaternary services		high-tech goods and services → ← increasing trade in high value goods and services		Increasing numbers of high-value and skilled industrial and office sectors

*In Phases 2–4, modern development in the global periphery tends to be concentrated in a relatively small number of countries, especially NICs such as India, South Korea and the Philippines.

Class activity

Study Figure 13.1.

1. Use terms such as dependency on the primary sector, terms of trade, branch plants, back offices, labour skills and comparative advantage to explain the evolution in the international division of labour.

2. Does Phase 4 offer better prospects for long-term development in the world periphery than Phases 1 and 2? Explain (see Figure 13.3 on page 89).

Phase 1 The traditional international division of labour

This began with the industrial revolution and continued until the 1960s. It involved the core economies of western Europe, and later the USA, in exporting manufactured goods to the global periphery. In turn, countries of the global periphery specialised in producing food and raw materials for the core economies. This phase was influenced strongly by colonialism (see Chapter 5).

In Phase 1, workforces in the core specialised in the secondary sector while the primary sector dominated in the global periphery.

Farm workers on a sugar cane plantation. Which phase of the international division of labour does this best illustrate? Why?

Phase 2 The new international division of labour

From the 1970s, in order to reduce production costs and remain competitive, many MNCs relocated branch plants to countries in the global periphery. At the same time, MNCs in core economies reduced their industrial workforces to concentrate on higher-value goods and services.

Low costs of labour allowed peripheral countries to specialise increasingly in branch plant production of industrial goods. This led to an increase in the world trade in merchandise goods.

The Microsoft Research and Development complex in Washington State, USA. Why are the headquarters and main research and development units of most MNCs still located in core countries?

85

Phase 3 The newer international division of labour

From the 1990s, a growing range of service functions has been relocated from core to peripheral countries. As with the earlier relocation of branch plants, these are basic service activities for which low labour costs are vital. Development of what are called **back-office functions** has created a new form of labour specialisation in the global periphery.

The core regions have responded to this relocation of both branch plants and a growing amount of back-office functions. As a result, their labour force specialises increasingly in high-tech industries and the growing quaternary sector.

Remember what is meant by back-office employment and the quaternary sector?

Indian employees work in one of the growing number of call centres that have been relocated to India. What phase of the international division of labour does this photograph represent? Explain.

Phase 4 The most recent international division of labour

A fourth phase of the international division of labour is just beginning, but its importance is expected to grow. In effect, MNCs are **outsourcing** or subcontracting an increasing number of key functions from their high-cost locations in core countries to locations in the world periphery (Figure 13.2).

This involves a significant number of high-value and skill-demanding jobs, such as research, chip design and financial analysis. These new jobs have been attracted to countries in the world periphery by a combination of push and pull factors:

In India, an engineer with an MSc designs computer chips for $1,000/month. A counterpart in the USA demands $7,000/month.

- Costs in core countries are very high, due mainly to high wages and salaries.
- Employees in peripheral countries, such as India and the Philippines, are becoming more educated and work for low wages. Improved communication systems also allow easy access to customers in the developed world.

Rather than well-educated workers in peripheral countries migrating to core economies, the work is now coming to the workers. This **reverses the long-established pattern of a brain drain from less-developed countries**. Do you think this is a positive step to promote development?

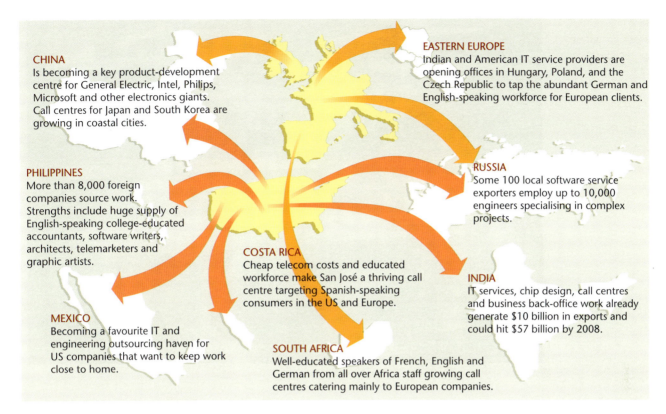

CHINA
Is becoming a key product-development centre for General Electric, Intel, Philips, Microsoft and other electronics giants. Call centres for Japan and South Korea are growing in coastal cities.

EASTERN EUROPE
Indian and American IT service providers are opening offices in Hungary, Poland, and the Czech Republic to tap the abundant German and English-speaking workforce for European clients.

PHILIPPINES
More than 8,000 foreign companies source work. Strengths include huge supply of English-speaking college-educated accountants, software writers, architects, telemarketers and graphic artists.

RUSSIA
Some 100 local software service exporters employ up to 10,000 engineers specialising in complex projects.

COSTA RICA
Cheap telecom costs and educated workforce make San José a thriving call centre targeting Spanish-speaking consumers in the US and Europe.

INDIA
IT services, chip design, call centres and business back-office work already generate $10 billion in exports and could hit $57 billion by 2008.

MEXICO
Becoming a favourite IT and engineering outsourcing haven for US companies that want to keep work close to home.

SOUTH AFRICA
Well-educated speakers of French, English and German from all over Africa staff growing call centres catering mainly to European companies.

Fig. 13.2 A world of outsourcing

LABOUR SUPPLIES IN THE GLOBAL PERIPHERY

The type of labour supply has been central to the attraction of branch plants and service functions to countries of the global periphery. **Four** features are critical for the evolving international division of labour.

1. **Costs of labour**. This is the most significant factor. Compared to core economies, costs of labour are low, and are therefore very attractive to companies wishing to reduce costs of production.
2. **Numbers of workers**. The high growth rate in the population, together with an underdeveloped economy, mean that a large supply of workers is available for basic tasks in most countries of the global periphery.
3. **Flexible workers**. The protection of workers' rights is generally weak. As a result, employers are often able to exploit their workforce through low wages, long hours of work and poor working conditions (see the boxed text on page 88).
4. **Education**. Initially this was not significant, especially for branch plants. By Phase 4 it had become very important. Therefore, countries such as India that have invested in educating their population are able to attract modern growth industries (Figure 13.3).

Women working in a textile factory in Vietnam producing jackets for export to Europe. What phase of the international division of labour does this best represent?

Because of these wages and work conditions we are able to buy the goods/services very cheaply. Is this a fair system?

The typical employee of a branch plant factory in the less-developed world is a woman between the ages of 16 and 26 years. In general, MNCs prefer women as they are more compliant than men in accepting working conditions that are significantly inferior in terms of security of employment, safety, shiftwork, wages and fringe benefits. These women usually work 20–30 per cent more hours per year than their counterparts in the labour force of developed countries – as much as 50 per cent in some cases. On the other hand, wages tend to be only 10–20 per cent of those paid in core economies. For example, experienced female labour can be paid as little as US$0.15 per hour in Indonesia; this figure rises to US$0.36 in Thailand, US$0.57 in El Salvador, and US$0.89 in Mexico. In some instances, these wages barely provide for the basic necessities of life. In spite of these working conditions, however, the productivity per worker tends to be only slightly below that of developed countries.

Class activity

Read the text shown above.

1. Describe the typical worker in a branch plant in the less-developed world.
2. Why do MNCs prefer to employ women?
3. If you are the owner of a company in western Europe and are concerned about your wage costs, why would Indonesia or Thailand be attractive for a new investment?
4. Do you think this type of employment is a good basis for development? Why?

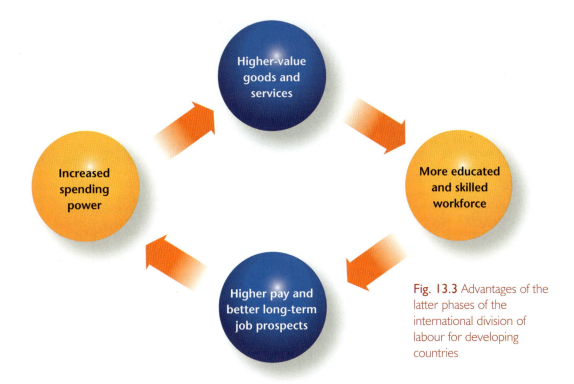

Fig. 13.3 Advantages of the latter phases of the international division of labour for developing countries

Class activity

Study Figure 13.3.
1. Explain the flow chart.
2. Why is this a better option for long-term development for peripheral countries?

Case Study: Nike

Nike, the US sports footwear and clothing company, provides a good example of globalisation and taking advantage of geographical differences between countries, such as the international division of labour. This is vital for their success.

Headquarters:	Bevertown, Oregon, USA
Annual sales:	$10 billion
Direct employment:	16,000 (mostly in global management, design and marketing)
Indirect employment:	500,000 (essentially production)
Original locations:	Core economies such as the USA, UK and Ireland. These production plants have all closed, since they were uncompetitive due to high labour costs.
Relocated to:	South-east Asia where Nike subcontracts production to companies located in over 30 countries. Manufacturing was focused originally in countries such as South Korea and Taiwan. However, as labour costs increased in these NICs, production has been relocated to even cheaper labour-cost locations in countries such as Indonesia, Thailand and China.

Nike shoe factory in Indonesia. Why should MNCs like Nike choose to locate factories in lesser developed countries? Do they provide any benefits for economic development in such countries?

Protests against what was seen as the company's 'sweatshop' working conditions in peripheral countries led Nike to join the **Fair Labour Association**. This global organisation promotes the ideal of a basic minimum wage to meet the daily needs of a worker and his/her family.

TEST YOURSELF AT
my-etest.com

SECTION 4 (CHAPTERS 14–18)
IRELAND AND THE EUROPEAN UNION

Ireland as a member of the European Union is part of a major trade bloc within the global economy. This has given rise to major forces for change within this relatively small and open trading economy. The impacts for Ireland have been both positive and negative. On balance, however, there seems little doubt that members hip of the EU has greatly helped Ireland's economic development.

This section has five chapters:

- Chapter 14   Trading Patterns of the European Union
- Chapter 15   Ireland's Trading Patterns within the European Union
- Chapter 16   The Common Agricultural Policy and its Impact on Ireland
- Chapter 17   The European Union and Ireland's Fishing Industry
- Chapter 18   The Common Regional Policy of the EU and Ireland

EU funding has been important for the modernisation of Ireland's transport network.

Roll-on/roll-off vessels such as these at the Ringaskiddy deep-water port in Cork allow heavy goods vehicles to move between Ireland, Britain and mainland Europe.

CHAPTER 14
TRADING PATTERNS OF THE EUROPEAN UNION

KEY IDEA!

The European Union is the world's largest trading bloc, with extensive internal and external trading links.

The countries of western Europe have a long and important tradition in world trade. This is linked historically to their roles as **colonial powers** which enabled them to dominate much of the world trade. Even following their loss of colonial empires, much of the growing world patterns of trade continue to focus on western Europe.

Refer to Chapter 5 to review the importance of colonialism and trade.

The EU forms the world's largest trade bloc. In 2002, the then 15 member states accounted for almost 40 per cent of all trade in goods and services. The development and patterns of EU trade, however, can be divided into two types:

● trade between its member states, called intra-EU trade
● EU trade with the rest of the world (extra-EU trade).

INTRA-EU TRADE

Prior to the signing of the Treaty of Rome in 1957, which established the European Economic Community (now called the European Union), trade between countries in western Europe was limited. This was due, in part, to the great rivalries that existed between countries, such as France and Germany.

Following the Second World War, Europe's economy was devastated and was in urgent need of redevelopment. The Treaty of Rome was central to this process and created a free trade area, a **Common Market** between member states. This meant the removal of barriers to trade which would encourage a growth in trade and prosperity for all member states.

Why is the River Rhine important for EU trade?

Growth in trade between member states has increased dramatically (Figure 14.1). Following slow growth in the 1960s, intra-EU trade expanded rapidly from the 1970s. By 2000, approximately 60 per cent of the US$2,500 billion trade of the EU flowed between its then 15 member states.

This growth and scale of intra-EU trade is due to:
● the Treaty of Rome (1957), which established the **principle of free trade** between member states
● a number of **enlargements**, which have increased the number of countries in the EU
● well-developed **transport and communication networks** to link member states
● the **size and prosperity of the EU market**, which encourages trade
● the creation of the **Single European Market** in 1993, which further helped trade flows within the EU.

Refer to *Our Dynamic World 1*, Chapter 27, to review enlargement of the EU.

The scale of intra-EU trade increased further in 2004 when ten countries from central and eastern Europe joined the EU. Prior to membership, these countries had already established important trade links to the EU (Figure 14.3). As they become more integrated in the EU and their economies develop, trade within the EU will grow faster.

It is estimated that enlargement to EU25 will increase intra-EU trade by 20 per cent.

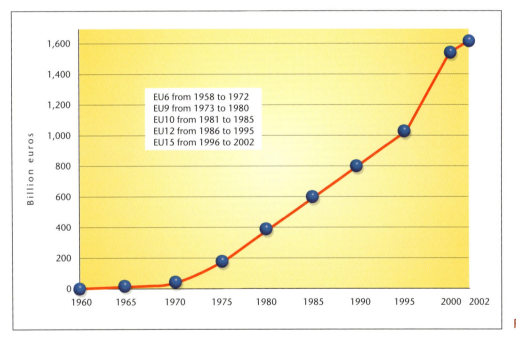

EU6 from 1958 to 1972
EU9 from 1973 to 1980
EU10 from 1981 to 1985
EU12 from 1986 to 1995
EU15 from 1996 to 2002

Fig. 14.1 Growth in intra-EU trade

Class activity

Review Figure 14.1.
1. Describe the trend in intra-EU trade from 1960.
2. Estimate the value of intra-EU trade in 2002 and compare this with the value when Ireland became a member state in 1973.
3. What impact does enlargement have on intra-EU trade? Explain your answer.
4. Do you expect the value of intra-EU trade to continue to rise? Explain.

Fig. 14.2 The five largest trading economies in the EU

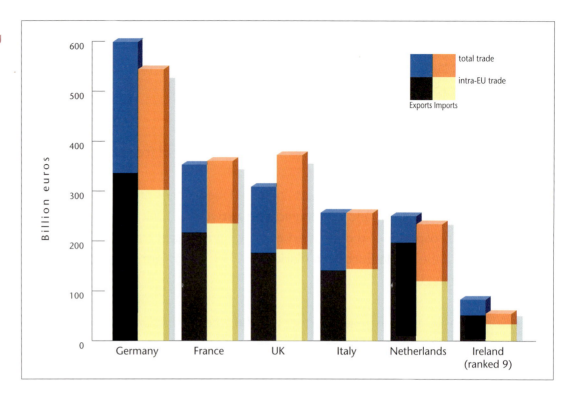

Can you suggest any reasons why Germany is so important for intra-EU trade?

Germany is the most important member state in terms of intra-EU trade (Figure 14.2). This country, together with the other top four trading countries of the EU, account for some 70 per cent of intra-EU trade.

Motorways are the most important form of transport for intra-EU trade. Can you suggest reasons why?

EU TRADE WITH THE REST OF THE WORLD (EXTRA-EU TRADE)

Due to its current size, levels of wealth and its historic role in global development, the EU has extensive trade links throughout the world (Figure 14.3).

While trade flows with candidate countries from central and eastern Europe grew strongly in the 1990s, **the USA and Japan are the most important trade partners for the EU**.

The EU also has important trade links with many developing countries. These can be traced to the region's former colonial empires located throughout the world e.g. the British and French colonial empires.

Rotterdam is the European Union's largest port. From the photograph suggest a product and a region of the world which are particularly important for Rotterdam's trade.

One of the most significant agreements to encourage trade and development with less-developed countries was the **Lomé Convention**. This originated in 1963 and established formal economic links between the EU and countries in **Africa, the Caribbean and the Pacific** (termed **ACP countries**). Most of these countries are former colonies of EU member states.

In an effort to encourage development, members of the Lomé Convention, now replaced by the Cotonou Agreement, are granted duty-free access for all industrial goods and for most agricultural commodities exported to the EU. While this has assisted some economic growth in ACP countries, they contribute a very small proportion of total EU trade (Figure 14.3). This is especially the case for sub-Saharan Africa, which includes some of the poorest countries of the world economy.

Remember the global trading triad? It is composed of the three global core regions that dominate world trade (see Figure 11.2 on page 74).

The Lomé Convention, which links the ACP and the EU, has been replaced by the Cotonou Agreement.

Why would duty-free access for goods exported to the EU help development in the ACP?

Remember branch plants, back offices and international division of labour in Chapters 9 and 13?

In addition to its formal linkages with ACP countries, the EU has experienced increases in trade with countries in **Asia and Latin America**. This relates to the attraction of these world regions (especially NICs) for branch plants and office-related activities.

The port of Fos near Marseilles in the south of France imports large amounts of coal and iron ore for its large steelworks. Oil is also imported through this port.

Fig. 14.3 Percentage share of the main partners for external trade in goods of the EU

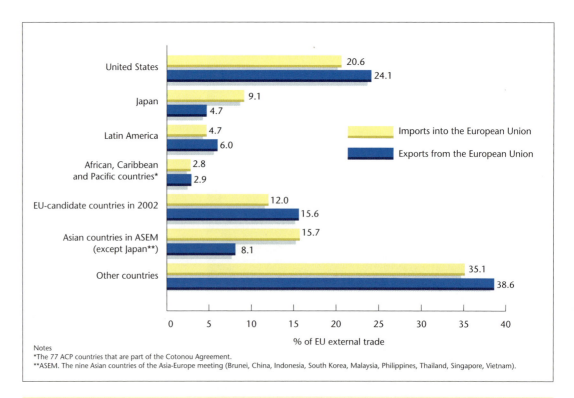

Notes
*The 77 ACP countries that are part of the Cotonou Agreement.
**ASEM. The nine Asian countries of the Asia-Europe meeting (Brunei, China, Indonesia, South Korea, Malaysia, Philippines, Thailand, Singapore, Vietnam).

Class activity
Review Figure 14.2.
1. What percentage of EU trade is with the other two countries of the global trading triad?
2. Which country is the dominant trading partner of the EU? Suggest some reasons for this.
3. What region has the lowest trade with the EU? Why?
4. Why has EU trade with countries in Asia and Latin America increased?

TEST YOURSELF AT

my-etest.com

CHAPTER 15
IRELAND'S TRADING PATTERNS WITHIN THE EUROPEAN UNION

KEY IDEA!

Membership of the European Union has had a major impact on the value, content and patterns of Ireland's trade.

A main aim of the EU is to promote free trade between its member states. In this, Ireland has done well. From being, in effect, a protected economy with limited foreign trade prior to the 1960s, Ireland is now one of the most open trading economies in the EU. **Since 1973, development of the country's economy has been driven largely by exports**.

Irish trade may be examined under three headings:
1. Changes in the value of trade
2. Changing patterns of trade
3. Changes in the make-up of trade.

Free trade within the EU has been crucial for the successful development of Ireland.

In 1959, a new industrial estate was opened adjacent to Shannon Airport. Special incentives, such as no customs charges on imported materials used by industries on the estate, boosted development and trade through the airport.

1 CHANGES IN THE VALUE OF IRISH TRADE

Prior to joining the EU in 1973, the total value of Irish trade was relatively small (Figure 15.1). EU membership resulted in a steady growth in trade, although the value of imports continued to exceed exports until the mid-1980s. This negative balance of trade was due to:

- high costs of imported energy supplies, especially oil
- Ireland's new branch plants involved the processing or assembly of imported raw materials and component parts.
- The mass-produced goods manufactured in branch plants for export did not generally have high value.

Fig. 15.1 Import and export trade of Ireland, 1972–2001

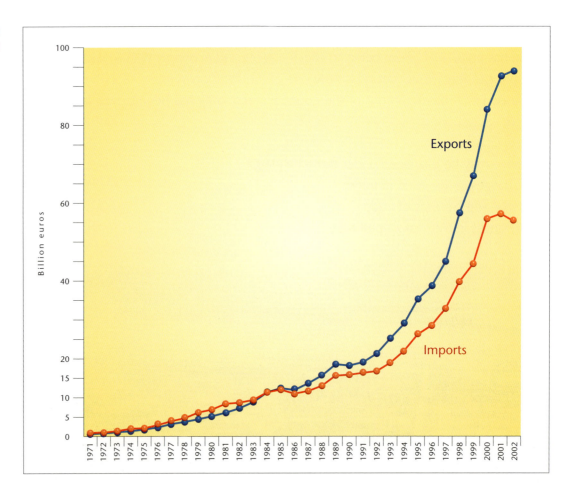

Class activity

Study Figure 15.1.

1. What is the main trend for both imports and exports?
2. Estimate the value of exports and imports in 2002 and compare with those for 1973.
3. Do these trends suggest membership of the EU has been good for Ireland's economy?
4. In what year did Ireland begin to show a positive balance of trade?
5. Briefly explain why Ireland's balance of trade is so positive in the 1990s compared to that of the 1970s.

From the mid-1980s, Ireland's trade increased rapidly. In addition, the value of exports began to exceed imports (Figure 15.1). These new trends became especially marked in the 1990s, and were due mainly to changes from a branch plant economy to the production of higher-value goods and services. The net result was a growing surplus in Ireland's balance of trade. This provided the Government with resources to invest in upgrading critical infrastructures e.g. telecommunications and education. **The strong trade performance in the 1990s is, therefore, a critical factor behind the success of the Celtic Tiger.**

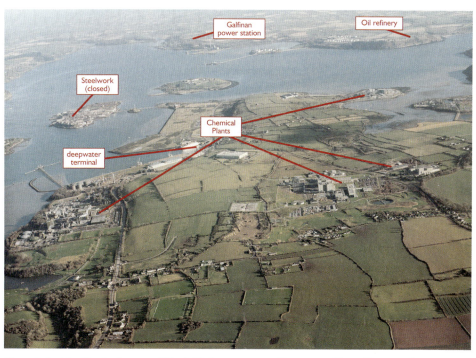

Many industries located around Cork Harbour process large quantities of imported raw materials e.g. oil refinery, chemical plants and steelworks (now closed).

For example, the manufacture of computer software requires few, if any, imports. Production depends mainly on the skilled and educated workers located in Ireland.

Suggest reasons why Dublin is Ireland's most important port for trading with the rest of the EU?

99

2 THE CHANGING PATTERN OF IRELAND'S TRADE

Historically, Britain has been Ireland's largest trading partner for both imports and exports (Table 15.1). An important reason for joining the EU was to reduce this level of dependence.

Until the early 1980s, Britain supplied at least half of Ireland's imports. However, as MNC branch plants located in Ireland began to source their imports from other EU and non-EU countries, so Britain's share of import trade declined. Despite this, *Britain remains Ireland's main import source*.

See for example Dell's import sources in Figure 10.5 on page 70.

Ireland's diversification of trade links points to its increasing involvement in the process of globalisation.

	1960	1972	1980	1990	2002
Percentage of exports to:					
UK	75	61	43	34	24
Rest of the EU	7	16	33	41	40
USA	8	9	5	8	18
Rest of the world	10	14	19	17	18

	1960	1972	1980	1990	2002
Percentage of imports from:					
UK	50	51	51	42	36
Rest of the EU	14	17	20	25	23
USA	10	8	9	15	15
Rest of the world	26	24	20	18	26

Table 15.1 Changing patterns of Ireland's trade

Changes in Ireland's export trade have followed the general trends for imports. However, the **scale of the changes in the geography of the country's export markets has been more dramatic** (Table 15.1).

Class activity

Study Table 15.1.

1. Which country traditionally dominated Ireland's import and export trade?
2. Is the UK a more important market for Ireland's imports or exports?
3. Describe trends in Ireland's trade with the rest of the EU. Did exports grow at a faster rate than imports?
4. Suggest reasons why Ireland's trade with the rest of the EU and the world has increased.

One of the main attractions of Ireland for MNC investment was its access to the large, prosperous and growing market of the EU. In effect, MNCs used Ireland as a **production platform** to gain free access to the EU. The net result has been a relative decline in Britain's dominant role and a marked increase in exports to the rest of the EU (Table 15.1).

As these MNCs began to adopt a more global marketing strategy, exports to both the USA and the rest of the world began to rise. The role of MNCs from the USA is of particular importance, since they account for almost two-thirds of Irish manufacturing employment in MNCs.

How do companies such as Liebherr, a German MNC located in Killarney, Co. Kerry, influence Ireland's changing pattern of trade?

3 CHANGES IN THE COMPOSITION OF IRELAND'S EXPORT TRADE

As part of Britain's colonial economy, Ireland's exports were dominated by agricultural goods. Despite gaining political independence in 1922, 50 years later, food and live animals still accounted for 43 per cent of the total value of Irish exports (Figure 15.2).

Membership of the EU, and its growing attraction for MNCs, however, led to a major change in the make up of Ireland's export trade. Despite the benefits to farming of the Common Agricultural Policy (CAP), the share of exports provided by farm produce fell dramatically to only 6 per cent by 2002.

The dominance of agricultural goods in Ireland's export trade was rapidly replaced by a range of manufactured goods (Figures 15.2 and 15.3). Multinational companies attracted to Ireland invested mainly in **international growth industries such as electronics and chemicals**. This was vital for the modernisation of the country's economy.

In addition, since Ireland is such a small market, **almost all MNC production is for export**. Membership of the EU is therefore crucial for the long-term success of these industries in Ireland.

For the influence of the EU and CAP on Irish farming, see Chapter 16.

Over 90 per cent of most chemicals and electronic equipment produced in Ireland is exported.

Fig. 15.2 Irish trade, by main sectors, 1972–2001

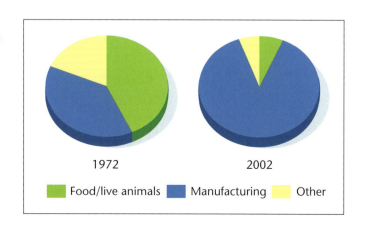

1972   2002

■ Food/live animals ■ Manufacturing ■ Other

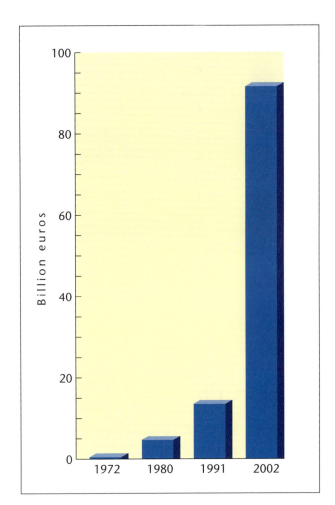

Fig. 15.3 Growth in the value of Ireland's manufacturing exports

Class activity

Study Figures 15.2, 15.3 and Table 15.2.

1. What was the most important export sector in 1972? Suggest reasons why.
2. What sector dominated Irish export trade in 2002? Why?
3. In what decade did the value of industrial exports increase the most? Why?
4. What growth industries account for the greatest value of Irish exports?
5. Are these industries dominated by MNCs or Irish-owned enterprises?

Table 15.2 Export trade for key international growth industries in Ireland, 1972–2002 (€ billion)

Industry group	1972	1980	1991	2002
Organic chemicals	14	112	1,471	17,122
Medical and pharmaceutical products	15	97	906	8,951
Office machines and automatic data-processing machines	NA	229	3,156	22,394
Electrical machinery	NA	307	989	9,223
Total four sectors	NA	745	6,522	57,690
Percentage of all manufacturing exports	NA	16	49	63

Growth of manufacturing exports was dramatic in the 1990s (Figure 15.3). Central to this were the large-scale developments in electronics and chemical industries.

By 2002, the four industrial groups shown in Table 15.2 accounted for almost two-thirds of the total value of all Irish exports. The majority of their exports are destined for markets in the EU and are produced by foreign-owned MNCs.

Membership of the EU has, therefore, been a major factor in the
● **growth**
● **changing geography and**
● **changing composition of Ireland's trade.**

CHAPTER 16
THE COMMON AGRICULTURAL POLICY AND ITS IMPACT ON IRELAND

KEY IDEA!

Since 1973, trends in Irish agriculture have been influenced mainly by the CAP.

> The Common Agricultural Policy is called the CAP.

In 1962, a Common Agricultural Policy (CAP) was introduced to guide the development of agriculture in the EU. Two of its main objectives were to:
1. increase production and productivity
2. provide a fair standard of living for all farmers.

These required two key policy decisions:
- Create a common tariff barrier (tax) around the EU to protect its farmers from cheaper imports.
- Establish an Agricultural Fund made up of Guarantee and Guidance sections to finance the CAP.

> The Intervention Price is the minimum guaranteed price farmers can expect for their output.

The **Guarantee Fund** maintains high prices for farm produce. It does this by intervening in the market to buy up any surplus output that would cause prices to fall from a high level, fixed each year by the EU. Prices are not allowed to fall below what is called the *Intervention Price*. To help reduce surpluses, the Guarantee Fund also subsidises exports to allow them to be sold in the world market.

The **Guidance Fund** provides money to help modernise farm buildings and machinery, and encourages increases in farm size. Through such supports, productivity levels are increased.

The CAP had a number of positive and negative impacts on EU farming (Table 16.1).

> Under the CAP, the more a farmer produced, the more income was made. This benefited larger, productive farmers and encouraged surplus output. Why?

Table 16.1 Some positive and negative impacts of CAP

Positive	Negative
High prices and guaranteed markets	Build up of large costly surpluses
Modernised farming	Larger and richer farmers benefited most
Increased productivity (intensive farming).	Environmental impacts such as soil and water pollution

Class activity

Can you suggest why each positive gain is linked to a negative consequence?

In 1992, the first of a number of reforms were made in the CAP. These have

- **reduced price levels** guaranteed to farmers
- **diversified farm activities** to provide additional income, especially for small farmers, e.g. farm-based tourism, forestry, craft industries such as cheese-making
- tried to make farmers **more competitive** in the EU and global market for food
- provided more **direct income support** for small-scale farmers
- stressed the role of the farmer in **protecting the environment**.

IRELAND AND THE CAP

In 1973, agriculture was vital for Ireland. It accounted for:

- 24 per cent of employment
- 43 per cent of the value of exports

and almost 50 per cent of population lived in rural areas.

Membership of the EU was **strongly supported by Irish farmers**. They saw many important advantages from the CAP:

- free access to a large and growing market
- reduced dependency on Britain's low-priced food market
- increased and guaranteed farm prices in the EU
- large transfers of money via high prices and funds to modernise farming.

Although agriculture has declined in importance since 1973, it still dominates the landscape over much of the country.

UNIVERSITY COLLEGE
Library
CORK

What does this cartoon suggest to you about Ireland's view of the CAP? Do you agree with this view?

Impacts of the CAP on Ireland

Farm output

Initially, farm output increased under the influence of higher guaranteed prices and access to the large EU market (Table 16.2). Farms were modernised and productivity increased.

Reforms of the CAP to reduce costly surpluses had a less positive impact on output trends in the 1990s. Furthermore, Ireland's important dairy sector was affected as early as 1984 when a **milk quota** was introduced to reduce milk output. This caused many small-scale dairy farmers, especially in the west of Ireland, to change emphasis in the 1980s from dairying to sheep (Table 16.2).

Table 16.2 Output in selected farm products and animal numbers, 1970–2002

	1970	1980	1990	2002
Product				
Cereals (000 tonnes)	1408	1711	1965	1964
Milk (million litres)	3629	5425	5268	5,039
Veal and beef (000 tonnes)	337	434	515	540
Animals (millions)				
Cattle of which:	5.9	6.9	6.8	7.0
Dairy	1.8	1.9	1.4	1.2
Sheep	4.1	3.3	8.5	7.2

Class activity

Study Table 16.2.
1. Which 10-year period showed the greatest increases for most products/animals? Why?
2. What has been the dominant trend for dairy cattle and milk from 1980 to the present? Explain this.
3. Contrast trends in sheep and dairy cattle. Suggest reasons for the differences.

Overstocking of sheep in upland areas has caused overgrazing and soil erosion. These environmental problems and lower prices led to a decline in sheep numbers in the 1990s (see Table 16.2).

Farm exports

The EU has provided a large and guaranteed market for Irish farm output. It resulted in **a major increase in the value of Ireland's food exports** (Figure 16.1). This has been an important contribution to the country's positive balance of trade (see Figure 15.1).

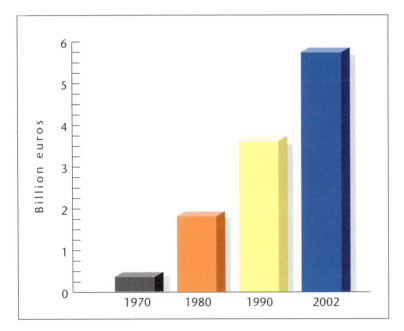

Fig. 16.1 Value of export trade in food and live animals, 1970–2002

Class activity

Review Figure 16.1.
1. Estimate the value of exports in 1970 and 2002.
2. Suggest reasons for this growth.

Farm income

Farm incomes have increased as they benefited from the high prices and market prospects in the EU. Despite this, they remain below the average industrial wage (Figure 16.2).

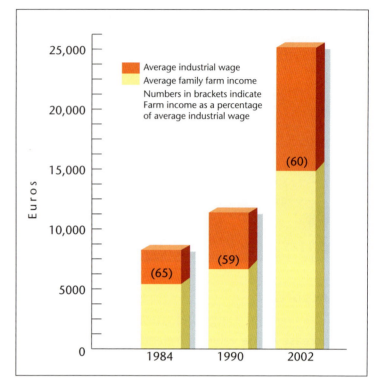

Fig. 16.2 Average annual family farm income and industrial wages

Table 16.3 Family farm income by main activity (€)

Region	1990	2002
Specialist dairying	15,800	28,100
Cattle rearing	3,100	7,800
Sheep	6,600	12,400
Tillage	18,000	21,900
Average all farms	8,400	14,900

Class activity

Study Figure 16.2 and Table 16.3.

1. Are family incomes higher than average industrial wages?
2. Suggest reasons why family farm incomes improved in the 1990s (see page 105).
3. Name the farm activities which provide the two highest and lowest incomes.
4. Which farm activities would you expect to dominate the west of Ireland?
5. Does Table 16.3 suggest a major division in farm prosperity between the west and east of Ireland?

TH E   C O M M O N   A G R I C U L T U R A L   P O L I C Y   A N D   I T S   I M P A C T   O N   I R E L A N D

- Under the CAP, the more farmers produced, the more they earned. This favoured large and efficient farms, especially in the Southern and Eastern region.
- Large numbers of farms, especially in the Border–Midlands–West region, are too small to produce enough output to earn a reasonable income.
- Inequality between small and large-scale farmers, and between the east and west of Ireland has, therefore, been increased by the CAP.
- Reforms to the CAP since the early 1990s, however, have provided more income support for small-scale farmers. This has helped improve their standards of living.
- Funding from the CAP is vital for Irish farmers. In effect, the vast majority of them depend on subsidies in order to survive.

> In 2002, 70 per cent of Irish farm income was provided by CAP subsidies. Does this suggest Irish farming is competitive?

> Specialisation means concentrating on a particular activity for which a farm has an advantage. This increases output and income.

Suggest regions of Ireland in which these farm environments are located. Which of these regions benefited most from the CAP? Explain why.

> Part-time farming has increased in importance for many small-scale farmers who want to stay on the land.

Numbers of farmers and farms

It had been expected that the benefits of CAP would slow down the **long-established decline in agricultural employment**. This, however, did not occur (Figure 16.3).

While farmers declined in number, the **average size of farms increased** (Figure 16.4). This was important to raise levels of income and productivity. Farms also became more specialised.

Fig. 16.3 Employment trends in agriculture, 1971–2002

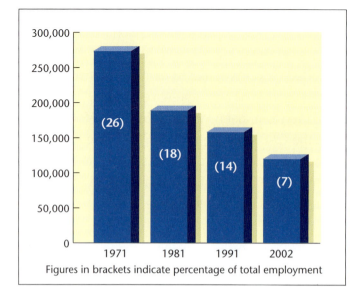

Figures in brackets indicate percentage of total employment

Can you suggest reasons for the continued decline in agricultural employment within the CAP?

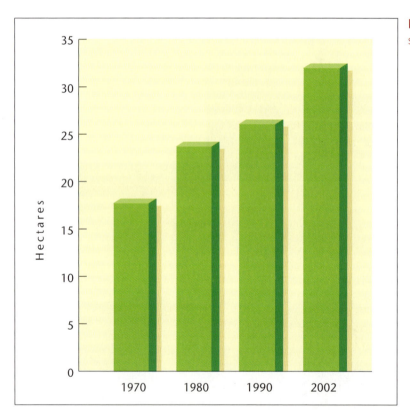

Fig. 16.4 Changes in average farm size, 1970–2002

Class activity

Study Figures 16.3 and 16.4.

1. Estimate the numbers employed in agriculture in 1971. How many agricultural jobs were lost by 2002?
2. Is the trend for farm size the same as for employment?
3. Is there a relationship between trends for employment and farm size? Explain.

Alternative farm enterprises, as shown in this photograph, have increased in importance, especially for smaller-scale farms. Why? Can you suggest any other enterprises farmers can use to increase their income?

Farming and the environment

To increase income under CAP, farmers were encouraged to use intensive farming methods. While output increased, this was often at the expense of the environment:

- Increased chemical fertilisers gave rise to soil and water pollution.
- Traditional field boundaries such as hedgerows and dry stone walls were removed to increase field size – this meant a loss of habitat for flora and fauna, and a decline in scenic quality.
- Over-stocking of the land – over-grazing and soil erosion, especially in uplands.
- Replacement of attractive and traditional farm buildings by modern structures.

Since the 1990s, the CAP has focused attention on reducing the environmental impact of intensive farming. Farmers are now offered subsidies to maintain and restore the quality of their farm environments. In Ireland, this works mainly through the **Rural Environmental Protection Scheme (REPS)**.

   The future of the majority of Irish farmers is now linked more to their abilities in managing a clean environment than their traditional role as producers of food.

The quality of the rural environment is crucial for Ireland's tourist industry.

Describe what this farmer is doing as part of the Rural Environmental Support Scheme. Why is this important for rural development in Ireland?

TEST YOURSELF AT

my-etest.com

CHAPTER 17
THE EUROPEAN UNION AND IRELAND'S FISHING INDUSTRY

 **KEY IDEA!**

The Common Fisheries Policy (CFP) of the EU has undermined development of the full potential of Ireland's fishing industry.

The shallow waters of the continental shelf, warmed by the North Atlantic Drift, are ideal conditions for the growth of **plankton**. This is the main food for fish.

For further information on Ireland's fishing industry, see *Our Dynamic World 1*, Chapter 18.

Ireland has control over a large area of the sea around its coastline. These **territorial waters** provide opportunities to exploit natural resources, such as natural gas and fishing (Figure 17.1).

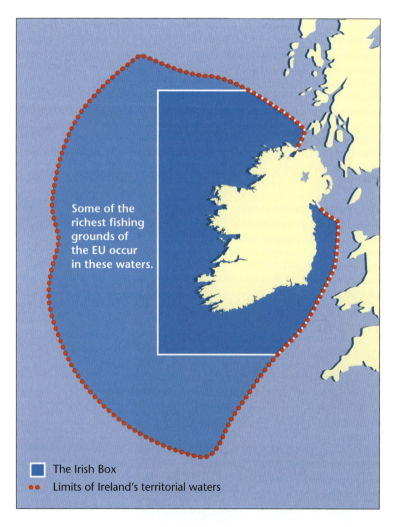

Some of the richest fishing grounds of the EU occur in these waters.

■ The Irish Box
●● Limits of Ireland's territorial waters

Fig. 17.1 Ireland's territorial waters

Despite its growth within the EU, the introduction of a **CFP in 1983 has restricted the full potential of Irish fishing**. A strong feeling has emerged that Ireland undervalued its rich fishing resources in return for ensuring continued benefits from the CAP.

Why do you think Ireland was more concerned with the CAP than a CFP?

Despite the introduction of TACs, overfishing is a major problem in EU waters. Why?

THE CFP AND ITS IMPACT ON IRELAND

Free Access to Ireland's Territorial Waters

This exposed the country's underdeveloped fishing industry to increased competition for its valuable fish resources. Spain, in particular, was seen as a major threat, but Ireland succeeded in preventing most of the Spanish fishing fleet from gaining access to the Irish Box (Figure 17.1).

One of the main concerns of the CFP is to prevent over-fishing and the destruction of this renewable natural resource.

The Irish Box and Spanish Fishing
Spain is the EU's largest fishing economy.

	Spain	Ireland
Number of vessels	16,500	1,330
Employment	64,600	6,000
Value of fish (€ million)	1,950	295
Weight of fish (000 tonnes)	1,080	320

In negotiations over Spain's membership of the EU, Ireland succeeded in greatly limiting the access of Spanish fishing boats into most of its territorial waters. This is called the Irish Box (Figure 17.1). Until 2002, only 40 Spanish fishing vessels were allowed to work in the Irish Box. These restrictions ended in 2003. Spain now wants full access to all of Ireland's territorial waters, while Irish fisherman want to preserve the restrictions of the Irish Box.

A part of Spain's large fishing fleet. Why should Ireland be concerned about allowing Spain full access to the Irish Box?

Fish Quotas

Despite possessing 11 per cent of the territorial waters of the EU, Ireland receives a quota of only 5 per cent of the TAC. This has restricted the development of the Irish fishing industry and has meant the creation of fewer jobs and less wealth, especially along the west coast. **Overfishing** of some valuable species, such as cod, has seen the TAC drop dramatically and has further affected the industry (Figure 17.2).

Fig. 17.2 Total allowable catch of cod in the Irish Sea

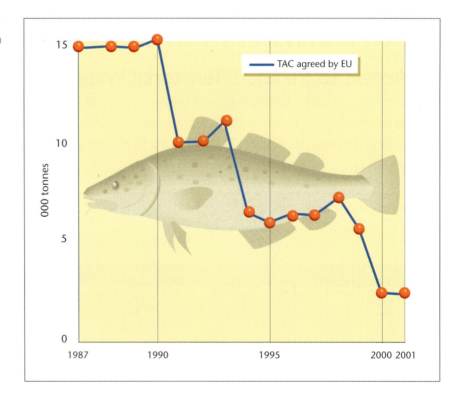

Free access and a small quota means that only 20 per cent of the value of fish caught in Irish waters is by Irish boats.

Class activity

Study Figure 17.2.

1. Estimate the total cod catch in 1987.
2. What has been the trend since 1987?
3. Does this trend suggest concerns with overfishing?
4. What impact does this trend have on fishing communities?

Total allowable catch (TAC)

In its attempt to prevent overfishing, the EU fixes a total allowable catch for each species of fish within its territorial waters. Each country is given a quota or percentage of the TAC.

Additional measures to prevent overfishing include minimum mesh size in nets to prevent catching young fish, and restricting or closing waters where fish stocks are under threat.

Fish Landings and Value

Despite restrictions, the total value and weight of fish landed have increased significantly (Figure 17.3). Without the quota, however, growth would have been even stronger.

Mackerel and herring are the most important catch by weight. The switch from overfished species such as cod to mackerel in the 1990s presented opportunities for a small number of fishermen. Operating mainly out of Killybegs, they invested in new and large fishing vessels and became known as **mackerel millionaires**.

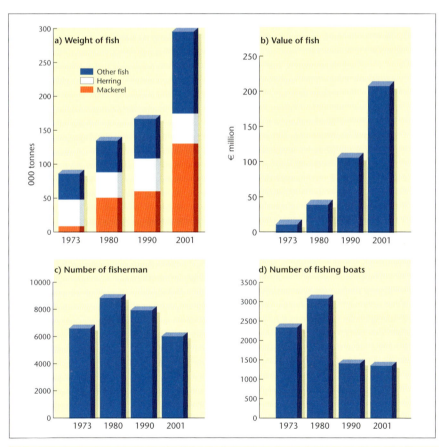

Fig. 17.3 Trends in Ireland's fishing industry

Class activity

Review the bar charts in Figure 17.3.

1. What is the trend for weight and value of fish landed in Ireland?
2. In what ways do trends for fishermen and boats differ from those for weight and value of fish landed?
3. Suggest reasons why the weight of fish landed has increased even though numbers of boats have declined.

115

Fish being unloaded at Killybegs, Ireland's largest fishing port. Why should fish-processing plants locate mainly at such ports?

Numbers of Fishing Boats and Fishermen

When Ireland joined the EU in 1973, higher prices and good market opportunities encouraged a growth in both fishing boats and fishermen. Since the introduction of the CFP and concerns with overfishing, however, these trends have been reversed (Figure 17.3).

Ireland's fishing fleet is now made up of a smaller number of larger vessels. These are able to catch more fish than the many smaller, but less efficient, boats they replaced. Rather than reducing problems of overfishing, this trend has increased them.

> Killybegs and Castletownbere are the two dominant fishing ports, accounting for one-third of the value of total catch.

Fishing Ports

With fewer but larger vessels, Ireland's fishing industry has become concentrated on a small number of major ports. These provide the essential services for such vessels. In addition, this gives rise to large and regular supplies of fish for the processing industry.

A REVISED COMMON FISHERIES POLICY

> Irish fishermen continue to argue for higher not lower TACs and greater protection of their fishing grounds from other member states, such as Spain.

The CFP was revised in 2003 to deal with the growing problem of overfishing. Greater stress has been placed on reducing total catch, and large areas have been closed for fishing. Growth prospects, therefore, for Ireland's fishing industry do not appear to be strong.

TEST YOURSELF AT

my-etest.com

CHAPTER 18
THE COMMON REGIONAL POLICY OF THE EU AND IRELAND

KEY IDEA!

Large transfers of Structural Funds from the EU have been vital for Ireland's development.

When Ireland joined the EU in 1973, it was the least prosperous of the, then nine, member states. Membership of the large and prosperous EU, therefore, held many advantages. Not least was the expectation of a large transfer of funds to help national development. In this context, the introduction of a **Common Regional Policy (CRP)** became a key benefit for Ireland.

Remember the expected benefits for trade, agriculture and industrial development?

Structural funds have been important in raising prosperity levels in Ireland. Despite a lot of investment, however, large areas remain underdeveloped in the Border–Midlands–West region.

In 1975, a CRP was introduced in the EU. This included the setting up of a **European Regional Development Fund (ERDF)**. The ERDF became one of four funds used by the EU to support development in problem regions of the EU. Collectively, these are known as **Structural Funds**.

Structural Funds have been vital for Ireland's development in the EU.

> ### Structural Funds
> These refer to four funds that are co-ordinated by the EU to support development in problem regions.
> - The **ERDF**, the main fund of the CRP, focuses on aiding industrial development and upgrading infrastructures, such as roads.
> - The **European Social Fund** (ESF) supports the training/retraining of workers and addressing problems of marginal communities, such as the unemployed and minority groups.
> - The **Guidance Section of the Agricultural Fund** aids the improvement of farm structures.
> - The **Financial Instrument of Fisheries Guidance** (FIFG) focuses attention on supporting the fishing industry and regions dependent on this sector.

Gross domestic product (GDP) is another way to measure national wealth. It is the total value of output produced in a country, but **excludes** net income from abroad. Contrast GNP on page 3.

To qualify for support under the CRP a region should have:
- a GDP per person below EU average
- a high dependency on agriculture or another declining sector, such as textiles or coal mining
- high and persistent unemployment.

Ireland qualified on all points and was designated as a single problem region. Between 1975 and 1988, the country received €3.2 billion in Structural Funds to help national development (Figure 18.1). This involved promoting new industrial activities and improving the country's transport systems, especially roads.

In spite of this, levels of prosperity in Ireland compared to those in the EU changed little (Figure 18.2). This was because
- **The amount of money was too little.**
- There was **poor use of funding.**

By 1988, Ireland's GDP remained less than two-thirds the EU average.

1989–99: A REFORMED CRP AND THE CELTIC TIGER

The CRP was reformed in 1988, and had a more positive impact on Ireland. In particular, it addressed the two problems that had limited its impact from 1975 to 1988.

Increased Structural Funds
The EU realised that if inequalities were to be reduced, Structural Funds had to be increased and targeted on the most depressed regions. This benefited Ireland.

Ireland's airports, but especially Dublin, have received a lot of money to improve their facilities. Why is this important for the country?

Ireland was designated an 'Objective 1' region. This recognised the severe problems of its peripheral location and under-developed economy. As a result, **the country was given high priority for Structural Funds.**

In the ten years to 1999, Ireland more than trebled its amount of Structural Funds (Figure 18.1). **This was a major source of money for the Government to invest in industrial development, modernise the infrastructure and improve the skills of its workforce.**

Objective 1 regions are the least developed in the EU. They have a GDP per person less than 75 per cent of EU average. See also *Our Dynamic World 1*, Chapter 15, page 188.

On a per-person basis, from 1975 to 1999, Ireland received more Structural Funds than any other problem region.

FÁS is the government body funded in part by the EU and charged with training and retraining workers in Ireland. Explain why FÁS, in places like Gweedore, Co. Donegal, is so important for regional and national development.

National Development Plans

To ensure Structural Funds are used effectively, Ireland now has to submit National Development Plants (NDPs) to the EU. These have forced the Government to plan strategically over a number of years, and are programmes for sustainable development.

Ireland's NDPs have invested heavily in three key and **interlinked** components of the economy:

- modernising the productive sectors, such as high-tech industry
- improving infrastructures, such as transport and communications to reduce the costs of a peripheral location
- increasing labour skills and flexibility to meet needs of modern growth industries.

During this period, Structural Funds and NDPs were central to the country's remarkable economic recovery. From being one of the poorest member states, Ireland's economic boom has transformed the country into one of the richest in the EU (Figure 18.2). This is the Celtic Tiger economy.

So far, Ireland has produced three NDPs for 1989–93; 1994–9 and 2000–06.

Successful development demands investment in all three components. Can you suggest why?

The City West High Technology Park near Dublin is a good example of interlinked planning. Use the photograph to explain this statement.

2000+: THE CRP AND A PROSPEROUS IRELAND

By 2000, Ireland's level of prosperity meant that it should no longer remain an Objective I region. In spite of this, the Government argued strongly that the country's peripheral location and large gaps in its infrastructure justified a continued inflow of Structural Funds.

The Government was successful, and Structural Funds continue to flow into the country. Two important differences, however, are to be noted.

1. The amount of Structural Funds has been reduced significantly (Figure 18.1). To compensate for this, Ireland has had to fund more of its development. So, while the third National Development Plan for 2000–06 involves an investment of some €52 billion, only €3.3 billion is from Structural Funds.

> Ireland's GDP per capita exceeds 75 per cent of the EU average, which defines an 'Objective I' region.

> More and more of Ireland's development will have to be self-financed.

Fig. 18.1 Structural Funds to Ireland, 1975–2006

Class activity

Review Figure 18.1.

1. How much Structural Funding did Ireland receive from 1975–88?
2. Which period shows the greatest inflow of Structural Funds?
3. Did this increase have a positive impact on Ireland?
4. Describe the trend in Structural Funds for Ireland in 2000–06. Can you explain this?
5. Is there any link between the trends for Ireland in Figure 18.1 and that in Figure 18.2? Explain.

---

2. Until 2000, the whole of Ireland was treated as a single region eligible for Structural Funds from the CRP. For the period 2000–06, however, the country has been divided into two planning regions (Figure 18.3). The Border–Midland–West (BMW) region had not benefited greatly under the Celtic Tiger and it has remained an Objective I region. This is important for the funding of its regional development.

Less justification could be made for the prosperous Southern and Eastern region to retain Objective I status. Here, Structural Funds under Objective I will be phased out until 2005. From that date, it will no longer receive priority funding.

Fig. 18.2 GDP per person in Ireland (EU = 100)

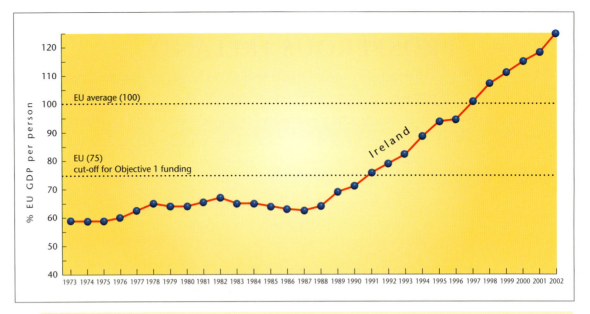

Class activity

Study the graph in Figure 18.2.

1. Estimate Ireland's GDP per person in 1973 as a percentage of the EU average.
2. How would you describe Ireland's trend in GDP per person from 1973 to 1988?
3. Do these trends change in the 1990s? Suggest reasons.
4. When did Ireland's GDP exceed 75 per cent of the EU average? What should this mean for an Objective 1 region?
5. In what year did GDP per person in Ireland exceed the EU average?

Fig. 18.3 Ireland's planning regions for Structural Funds

Until 2005, the Southern and Eastern region is referred to as Objective 1 in transition.

Class activity

Divide the class into two groups. One group is responsible for planning development in the BMW region, the other in the S and E. Debate **three** investment priorities to promote regional development. Are they the same for each region?

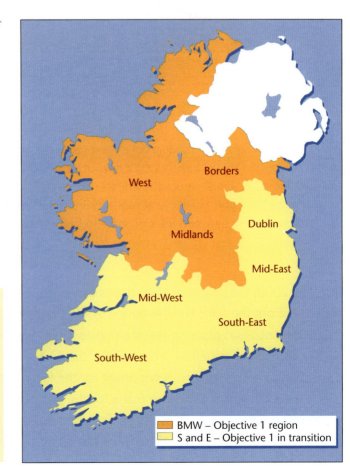

NDPs and Structural Funds have been a success story for Ireland. The country's productive sectors, infrastructure and human resources have all benefited enormously. This is essential for a peripheral economy like Ireland to compete in a future EU of 25 or more member states.

By 2006, Ireland will contribute more to the EU budget than it will receive!

Despite increasing prosperity in Ireland, many people continue to live in poverty. Why is Social Funding important to help combat poverty?

Social funding

In addition to regional inequalities, significant social inequalities also exist in Ireland. **Social exclusion** is the term used to define the process whereby some people are excluded from the benefits of development. A key aim of the NDPs, therefore, is to replace social exclusion by **social inclusion** to ensure that all people can benefit from increasing prosperity. In this, the **European Social Fund (ESF)** has a vital role by providing money for:

- reducing unemployment through **training and retraining schemes** (especially for long-term and youth unemployment)
- providing more **affordable housing** for the poorest members of society
- helping to **integrate minorities**, such as Travellers and refugees into society
- **gender equality**, by improving work opportunities for women, e.g. crèches, training programmes.
- **community support schemes** to help improve the quality of life for people living in disadvantaged areas, e.g. high-density working-class housing areas in inner cities.

Class activity

Do you think that social funding is as important for Ireland's development as ERDF support to improve the country's transport systems and to attract new industries? Explain.

TEST YOURSELF AT
my-etest.com

SECTION 5 (CHAPTERS 19–25)
ECONOMIC ACTIVITIES HAVE AN ENVIRONMENTAL IMPACT

There is a strong relationship between economic development and the use of the earth's natural resources. Both renewable and non-renewable resources provide people with the energy supplies and raw materials required for development, as well as basic needs such as clean water to drink and fresh air to breathe.

The scale and rapid increase in economic development throughout much of the world, however, has placed an increasing strain on the earth's resources and natural environments. Pollution of the environment, in particular, is a growing threat to further development and our quality of life. It also causes increased conflict between those people who favour using the environment for short-term economic growth and those who support longer-term sustainable development.

This section is divided into seven chapters:

- Chapter 19   Renewable and Non-Renewable Resources
- Chapter 20   The Environmental Impact of Burning Fossil Fuels
- Chapter 21   Renewable Energy and the Environment
- Chapter 22   Environmental Pollution
- Chapter 23   Sustainable Economic Development and the Environment
- Chapter 24   Economic Development or Environmental Protection: A Cause of Local Conflict
- Chapter 25   Economic Development or Environmental Protection: Global Concerns

The rich natural environment of tropical rainforests.

Chronic industrial pollution.

The Women's Environmental Network protest against traffic pollution in London.

CHAPTER 19
RENEWABLE AND NON-RENEWABLE RESOURCES

KEY IDEA!

The use of renewable and non-renewable resources, and especially energy resources, has been vital for economic development.

Natural resources include mineral deposits (e.g. iron ore, copper), fossil fuels, such as coal and oil, water, air and soil. They can be subdivided into renewable or non-renewable resources.

Renewable resources *can be replaced* through natural processes. If they are used at a faster rate than they are replaced these resources can be run down to levels at which they become of little or no use to people.

Non-renewable resources are available only in a *finite amount*. They cannot be replaced once they have been used. The availability of these resources can be extended through careful planning, e.g. recycling and more efficient use.

Natural resources are used by people to meet their needs, such as food, and as inputs for industrial production, e.g. iron ore.

Class activity

Which of the following natural resources are renewable, and which are non-renewable? Explain the reasons for your choice:

- Forestry _____
- Oil _____
- Coal _____
- Fish _____
- Soil _____
- Iron ore _____

ENERGY RESOURCES

Energy resources have been, and remain, vital for economic development. For example, the industrial revolution, which was centred on western Europe throughout the nineteenth century, was based on the cheap and large-scale availability of *coal*. Coalfields, therefore, became major industrial regions, such as the Ruhr and Sambre Meuse. Since the Second World War, however, coal has been replaced increasingly by other energy sources, especially oil and natural gas.

The European Union and Energy Resources

The EU has a large and varied range of energy resources. This is reflected in its increasing levels of production (Figure 19.1).

The main sources of energy production from within the EU have, however, changed significantly (Figure 19.2). While **fossil fuels** continue to dominate energy production, oil and natural gas have replaced coal as the main sources of supply. This has been due primarily to the discovery and exploitation of large **oil and gas resources in the North Sea**.

Production of **nuclear power** has increased rapidly since the 1970s. It has become especially important for some countries which lack alternative domestic energy resources (e.g. Belgium, following closure of their coalfields). **Renewable energy** sources, such as hydro-electric power, are important in Alpine Europe and Scandinavia.

Consumption of energy has also increased within the EU (Figure 19.1). This reflects the region's continued economic development. Production from within the EU is unable to meet this demand. As a result, almost half of the region's energy needs have to be imported.

> Coal provided 90 per cent of energy supplies in 1950 but only 14 per cent in 2000.

Britain and Norway, in particular, have benefited from the major development of North Sea oil and gas. Can you suggest a reason why? (Hint: Look at a map of the North Sea Basin.)

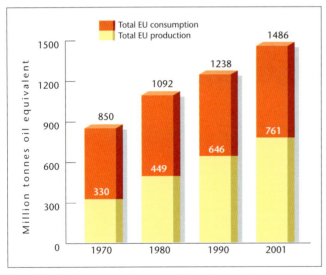

Fig. 19.1 Production and consumption of energy in the EU, 1970–2001

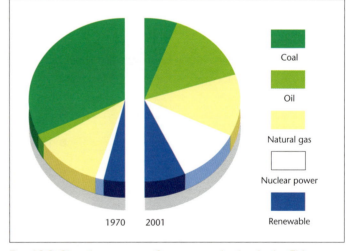

Fig. 19.2 Changing sources of energy production in the EU

Class activity

Study Figures 19.1 and 19.2.

1. Describe the trends for energy production and consumption.
2. Is the EU self-sufficient in energy supplies?
3. Describe the main changes in production of energy within the EU between 1970 and 2001. Can you suggest reasons for these changes?

National Energy Resources

Countries such as Britain, which possess a large range of domestic energy resources, have important advantages for economic development.

In contrast, where national energy resources are poor, as in Ireland, energy has to be imported. This adds to production costs and exposes the economy to the risks of disruption to supplies of energy, as in the oil crises of the 1970s.

Different countries in the EU depend on different resources to meet their energy needs (Figure 19.3). Most, however, are not self-sufficient in domestic energy supplies and are dependent upon imports.

List some advantages for a country which has a good range of energy resources.

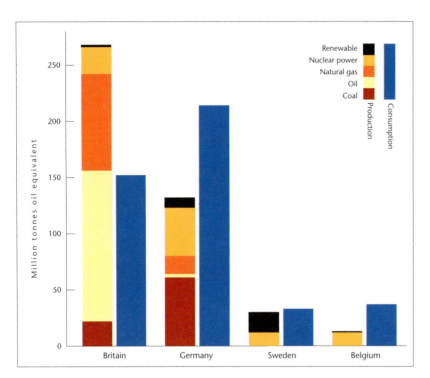

Fig. 19.3 Energy production and consumption in selected EU countries

A large open-cast lignite mine in east Germany. While this is an important energy source, does it appear to be environmentally friendly?

Class activity

Study Figure 19.3.

1. Which country is the largest producer of energy in the EU and a net exporter of energy? Can you explain this?
2. Which country is most dependent on renewable energy? What type of renewable energy?
3. In which country does nuclear power dominate the domestic production of energy? Why?
4. What country is the most dependent on coal? Name the most important coalfield in this country.

France has the largest nuclear power industry in the EU. This energy sector was developed since France now has few alternative sources of energy.

The Shannonbridge peat-fired power station. Note the extensive area of raised bog mined to provide peat for this power station. Why are such power stations important for Ireland?

Ireland and Energy Resources

Ireland has relatively few energy resources (Figure 19.4). On gaining independence from Britain, the Government attempted to increase the use of its limited sources of energy supply. This has involved both renewable and non-renewable forms of energy.

Through Bord na Mona, the Government exploits large areas of bogland. These provide a low-grade fossil fuel in the form of **peat**. This is used for domestic heating and supplying three peat-fired power stations.

Renewable energy supplies involve **hydro-electric** *power stations* on some of Ireland's major rivers. More recently, there has been growing interest in trying to harvest **wind power**, given the strong and regular wind flows that cross the country (see page 137).

Fig. 19.4 Energy resources of Ireland

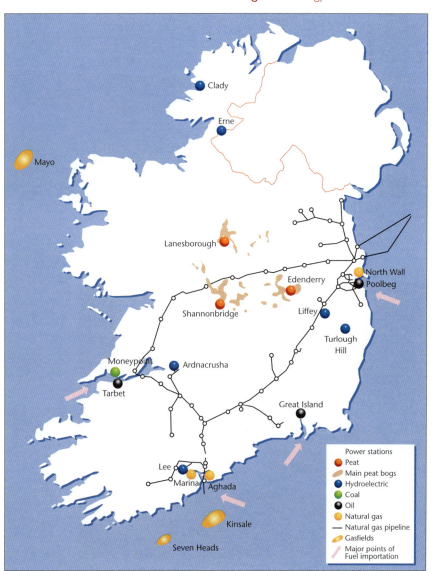

Offshore exploration for oil and gas in the 1970s led to the discovery and exploitation of a significant *gas field* off the Cork coast. This gas (and imported gas from Britain) is redistributed through an extensive system of gas pipelines. More recently, a large gas field has been found off the Mayo coastline. This should have major implications for this peripheral region.

Consumption of energy has increased rapidly since the late 1950s (Figure 19.5). Since domestic production accounts for about only 15 per cent of the country's energy needs, Ireland is very dependent on imports of energy to promote national development.

Natural gas accounts for two-thirds of Ireland's domestic production of energy.

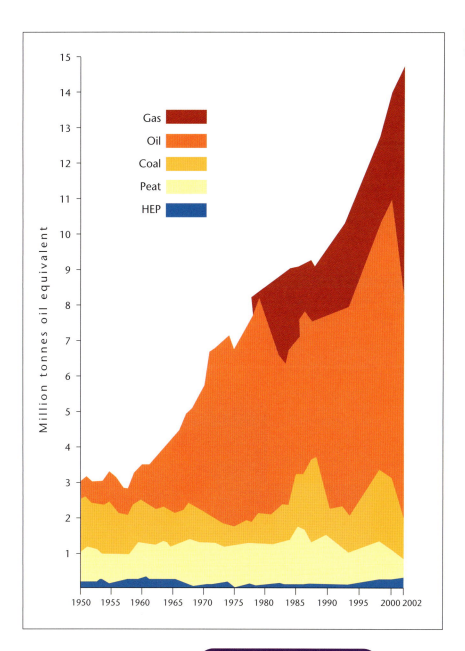

Fig. 19.5 Pattern of energy consumption in the Republic of Ireland

Class activity

Study Figures 19.4 and 19.5.

1. Why are gas pipelines important for Ireland's energy supply?
2. Why are Ireland's largest power stations located near major ports?
3. Describe the trends in energy consumption. Suggest why they increased after the 1950s.
4. Which energy source has dominated consumption since the 1960s? Is this a domestic resource?
5. Which two fossil fuels are produced in Ireland? Contrast their locations of their main sources and trends in consumption.

CHAPTER 20
THE ENVIRONMENTAL IMPACT OF BURNING FOSSIL FUELS

KEY IDEA!

Acid rain and declining air quality are two important consequences of burning fossil fuels.

Fossil fuels include coal, lignite or brown coal, peat, oil and natural gas. Since the industrial revolution, these natural resources have dominated the energy markets of developed urban and industrial regions. To release their stored energy for human use, fossil fuels have to be burned. This takes place in homes (domestic heating), factories (e.g. iron and steel plants) and especially in power plants which burn fossil fuels to produce electricity.

Burning fossil fuels has increased enormously over the last hundred years and has had a major impact on the environment. This chapter outlines *two* impacts: acid rain and air quality.

Burning fossil fuels also releases large amounts of carbon dioxide (CO_2). They are called greenhouse gases and give rise to global warming (see page 145).

ACID RAIN

Burning fossil fuels releases large quantities of sulphur dioxide (SO_2) and nitrogen oxides (NO_x) into the atmosphere. These gases react with water vapour to form weak acids. Precipitation then falls to earth as *acid rain*.

The Geography of Acid Rain

Remember acid rain from your Junior Certificate studies?

All parts of Europe are affected by acid rain. The regions with the largest fallout from acid rain form an east–west zone stretching from Britain to Poland (Figure 20.1). This includes most of Europe's major urban-industrial centres and are the sources for most SO_2 and NO_x pollution.

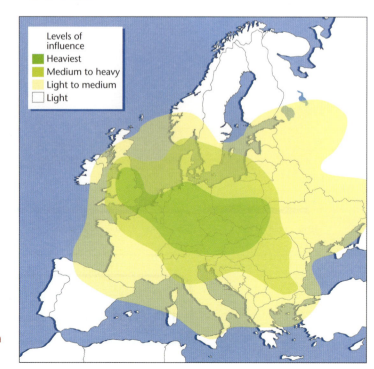

Levels of influence
- Heaviest
- Medium to heavy
- Light to medium
- Light

Fig. 20.1 Levels of influence from acid rain

Prevailing winds can cause acid rain to fall far from its source region. As a result, large areas of peripheral Europe which do not burn significant amounts of fossil fuels are affected by acid rain (Figure 20.1).

Impacts of Acid Rain

The impacts of acid rain are varied, affecting both the natural and human environments, as well as human health.

- **Forests are damaged** as acid rain increases the vulnerability of trees to pests and disease. Tree growth is stunted, leaves become discoloured and drop early. In extreme cases, the tree dies. An estimated one in four of Europe's trees are affected by acid rain (Figure 20.2).

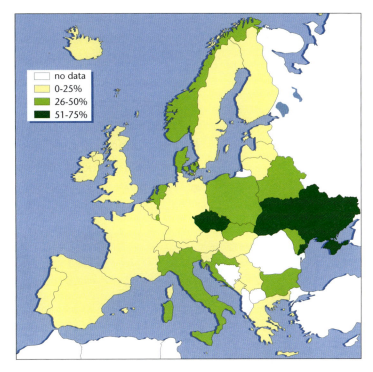

no data
0-25%
26-50%
51-75%

Fig. 20.2 Proportion of trees in Europe damaged by acid rain

We turned off the road into a rough track. 'Climb up there,' he said, 'then you'll understand what acid rain is all about.' On the way up I nearly fell into a deep gully gouged out of the mountainside by water rushing down the treeless slopes. From a high ridge I saw a forest of stark, grey, dead tree trunks extending as far as the eye could see. The feeling of desolation was overpowering. If the chimney smoke has had this effect on trees, what could it be doing to human health, I wondered.

A description of the effects of acid rain in south-west Poland
National Geographic Magazine, June 1991.

Class activity

Study Figure 20.2 and read the extract from the *National Geographic Magazine*.

1. What proportion of Poland's forests are damaged?
2. What parts of Europe show the greatest proportion of forests damaged by acid rain?
3. Explain the high proportion of damaged forests a) in Norway and b) in central/eastern Europe.
4. Compare Figures 20.1 and 20.2. Is there a connection between damage to trees and acid rain?

Eastern and central Europe suffer especially badly from acid rain. Reliance on brown coal/lignite which has very high sulphur content plus poor environmental controls of dirty, heavy industries under Communism meant that SO_2 pollution was very high. Since 1990 many heavy industries have closed down and pollution levels have declined.

- **Lakes and rivers** have their acid levels increased. This affects both fish life and plant growth. In southern Norway and Sweden, the fish populations in a majority of lakes have declined. Lime is being deposited in the most badly damaged lakes to counteract the influence of acid rain. This is an expensive process.
- **Leaching of soils** increases. This removes essential nutrients for plant growth and also causes toxic minerals to enter rivers and lakes.
- Many **important buildings and statues are damaged** as acid rain increases the weathering process, especially of limestone, marble and some sandstones.
- High levels of SO_2 and NO_x in the air can **cause health problems**, especially for those with respiratory problems.

Forest destruction caused by acid rain.

Weathering of stonework increases through acid rain, causing significant damage to many historic buildings.

One of the most severe winter smogs occurred in London in December 1952. Over 4,000 deaths, through respiratory and heart problems, were linked to the smog. This led to the 1956 Air Pollution Act which quickly helped clean up air quality in British cities.

SMOKE POLLUTION AND SMOKE-FREE ZONES

The burning of solid fuels, such as coal and peat, not only release greenhouse gases into the air, but also large quantities of solid particles or *smoke*. Historically, this problem was greatest in major urban areas where heavy industries and large numbers of people burned coal as their main form of energy and domestic heating. So, in the absence of strict air pollution controls, the larger the city, the poorer the quality of air.

Smoke and air pollution are generally highest in winter and can cause *winter smogs*. These occur especially in combination with low temperatures (which increase the burning of coal for domestic heat), and low wind speeds and temperature inversions (which prevent the dispersal of the pollution).

Case Study: **Dublin**

Throughout the 1980s, the main air quality problem in Ireland was the occurrence of winter smog. This resulted from the widespread use of coal, particularly in Dublin (Figure 20.3).

By the mid-1980s Dublin was producing an average of 55 tonnes/km²/year of smoke-based pollution. This was six times the average amount produced in London. It was claimed that *Dublin was probably the most smoke-polluted city in Europe.*

In 1987, some 20 years later than in Britain, Ireland introduced its Air Pollution Act. As a result, in 1990 a 'coal ban' was introduced in Dublin which prevented the marketing, sale and distribution of bituminous coal in the Dublin area. This created a smoke-free zone for the capital city.

One of London's infamous smogs prior to the introduction of smoke-free legislation.

Bituminous coal gives off relatively high levels of smoke and SO₂ when burned. This has been replaced by smokeless coal. Access to natural gas by pipeline from the Kinsale Head gasfield has also helped reduce levels of smoke pollution.

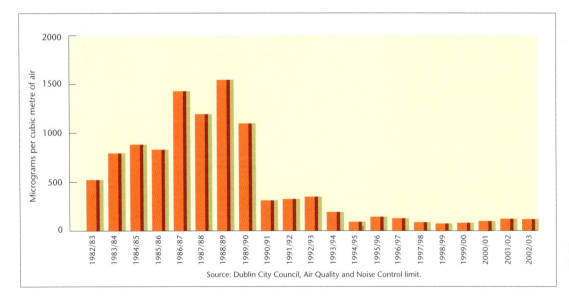

Source: Dublin City Council, Air Quality and Noise Control limit.

Fig. 20.3 Highest daily black-smoke levels in Dublin City, 1982–2003

Class activity
Study the graph in Figure 20.3.
1. Describe and account for the trend in smoke pollution in the 1980s.
2. In what year was the highest level of smoke pollution?
3. In what year was there a dramatic fall in levels of smoke pollution? Why?
4. What has been the trend in levels of smoke pollution since the early 1990s?

The ban on using 'dirty' coal resulted in a dramatic fall in levels of smoke pollution in Dublin (Figure 20.3). In effect, the city is now almost smoke-free. This has had a major impact on air quality within the city. Medical research has also shown that the ban has resulted in some 116 fewer respiratory deaths and 243 fewer heart-related deaths per year in Dublin.

The ban on the sale of bituminous coal and the creation of smoke-free zones now covers 16 urban centres in Ireland, including all five main cities. All have shown a considerable reduction in smoke pollution and improvement in air quality.

Smog over Ballymun before the 'coal ban' was introduced in Dublin in 1990. Has air quality improved in the city since that date?

TEST YOURSELF AT
my-etest.com

CHAPTER 21
RENEWABLE ENERGY AND THE ENVIRONMENT

KEY IDEA!

Renewable energy sources are effectively a clean source of electricity, although there are some environmental impacts.

Sources of renewable energy include biomass, geothermal, solar, tidal, wave, water and wind. Most have been used on a small scale for thousands of years. Today, about 20 per cent of global energy is supplied by renewable energy, mostly in the developing world. In developed countries, however, renewable energy sources generally supply only a small proportion of total energy consumption, e.g. 6 per cent in the EU and 2 per cent in Ireland.

Since the 1980s, there has been increasing interest in renewable energy.

- Improved technologies allow for more efficient production of electricity from renewable resources.
- With effective management, they will not be exhausted (as fossil fuels do) and so can provide a long-term solution to global energy needs.
- They are generally clean sources of energy and do not pollute the environment.
- There is greater government support for renewable energy.

In the ten years to 2010, the EU plans to double to 12 per cent renewable energy's share of its total energy supply. This will help meet its promises to reduce greenhouse gas emissions under the Kyoto Protocol (see page 147).

This chapter reviews the development and environmental impact of water and wind power.

> **What do you understand by the terms biomass, geothermal and solar energy?**

WATER POWER

The use of flowing water to create hydro-electric power (hep) is seen by many as an important alternative energy source to fossil fuels. In spite of this, several problems prevent its growth and greater use:

1. Most of the **best sites** for hep have **already been developed**.
2. **High costs of construction** of the dams and reservoirs of large hep stations – this is especially important for developing countries.
3. Large hep stations can cause **some environmental problems**.

> **Remember how hep is generated, from your Junior Certificate studies?**

The larger the hep scheme, the greater the problems become. Why?

These include:

- loss of land due to flooding to create the reservoir
- resettlement of people displaced by the reservoir
- changing environmental qualities for plant, animal and fish life
- reduction in the visual quality of the landscape caused by building complex and transmission lines to distribute electricity.

Ireland and Hep

Water power was the first domestic energy source developed to generate electricity in Ireland after independence. The first of the country's hep schemes was opened at Ardnacrusha in 1929. In the 1930s, it became a vital source for extending electricity supplies throughout rural Ireland.

Most of Ireland's hep capacity has now been tapped. It is unlikely that any large new hep station will be built. By 2002, only 2 per cent of Ireland's electricity was provided by hep.

The Ardnacrusha hep site on the River Shannon. Do you think the environmental problems linked to the development of hep were an important consideration for this project?

WIND ENERGY

Wind is one of the most underdeveloped renewable energy resources. Many countries are now, however, investing in wind power as an effective and clean source for generating electricity.

Preferred sites for wind farms cover large areas, have low population densities, and are exposed to regular winds. Coastal lowlands with onshore winds, or hill/mountain-tops, provide such locations. Offshore areas are also becoming attractive options for wind farms.

Although a clean source of energy, there are some **environmental impacts** linked to wind farms:

- noise caused by the rotating wind blades
- the visual impact of the large wind turbines – especially as they are often located in attractive uplands/coastal lowlands
- the large areas covered, although farming can be continued around the turbines
- birdkill.

Wind Power in Ireland

The focus of Ireland's recent national policy for renewable energy is *wind*. This is linked to the country's large and regular flow of onshore, south-westerly winds.

Despite good site conditions, only 1.5 per cent of Ireland's electricity in 2002 was produced from wind farms (Figure 21.1). It is planned, however, to increase the number of wind farms and their importance for energy supply. If targets are met, 7 per cent of Ireland's electricity will come from wind power by 2005.

> A wind farm is a group of wind generators which feed electricity into the national grid.

> Western Denmark and northern Netherlands are important regions for wind power. Why?

A wind farm in Co. Donegal. Why are such sites suitable for wind farms? Do you think they are unattractive features in the landscape?

Fig. 21.1 Wind farms in Ireland

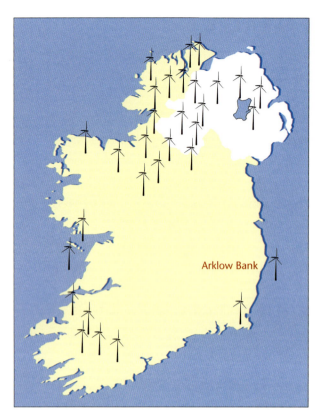

Class activity

Study Figure 21.1.

1. How many wind farms are located in the Republic of Ireland.
2. Describe and account for their location pattern.
3. Summarise the impacts of wind farms on the environment (refer also to photographs on pages 137 and on this page).
4. Do you think the benefits of wind power outweigh their local disadvantages?

In 2002, the Irish Government gave planning permission to begin construction of the world's largest offshore wind farm. This is on the Arklow Bank, a sandbank 7 km off Arklow, Co. Wicklow. Currently, eight wind turbines are under construction, but it is planned to build 200 in total. These will increase four times the present electricity output from wind. It will serve the needs of some 500,000 people along the east coast and will cost €630 million to construct.

Service boats are dwarfed by the giant turbines already in place (2003) on the Arklow Banks wind farm. Do you think such offshore locations are suitable sites for wind farms? Why?

For renewable energy sources to become a major source of power, and thereby reduce environmental pollution from fossil fuels and nuclear power, far greater government investment is necessary. Also, we as consumers will have to change our attitudes to energy supply. Are *we prepared to pay more for our energy in order to have a cleaner environment*?

What are your thoughts on this subject?

TEST YOURSELF AT
my-etest.com

CHAPTER 22
ENVIRONMENTAL POLLUTION

KEY IDEA!

Pollution knows no boundaries and can impact on the environment at local, national and global scales.

Since the industrial revolution, improved technologies and a rapidly increasing population have placed huge pressures on natural resources. Not only are non-renewable resources depleted, but also the by-products of development create major problems of *pollution*.

Furthermore, **pollution knows no boundaries**. As a result, pollution is rarely confined to the area in which it originates.

Pollution means the release of substances, primarily from human activities, in amounts which cannot be absorbed naturally by the environment.

Each person in Ireland generates a heap of rubbish similar to this every year.
Does this create problems for our environment?

Remember **acid rain** in Chapter 20. Is this an example of 'pollution knows no boundaries?'

POLLUTION AT A LOCAL/NATIONAL SCALE

Waste Disposal in Ireland: What a load of rubbish!

There has been a massive increase in the build-up of solid waste, or rubbish, in Ireland. As our levels of prosperity have increased, we have become a 'throw-away society'.

The preferred option, to date, for waste disposal in Ireland has been to use the 50 or so local authority landfill sites (rubbish dumps) located around the country. Most of these sites, however, are nearing their capacity.

In addition, most local communities are strongly opposed to any planning application to enlarge or open a new landfill site in their locality.

This is often referred to as the NIMBY attitude. As a result, there is an urgent need to find an alternative solution for the country's **growing mountains of waste**.

What is meant by NIMBY? Identify **three** pollution problems linked to landfill sites.

Incineration of Waste: a burning issue!

This is a recent means of disposing of solid waste. It involves burning waste in large incinerators. The process, however, is very controversial between supporters and those who oppose it, especially in terms of environmental pollution (Table 22.1).

In favour of incineration	Opposed to incineration
● Takes up little space	● Air pollution concerns
● Deals with large volumes of waste	● Incineration releases dioxins into the air which can cause cancer
● Generates heat which provides an additional source of energy	● Residue ash can contain toxic materials which have to be disposed of
● Technology ensures only limited pollution	

Table 22.1 Arguments for and against incineration

At least six major incinerators are planned to deal with waste on a regional basis in Ireland. For example, one planned for Poolbeg in Dublin could treat 25 per cent (400,000 tonnes) of the city's waste, and generate power for 35,000 homes.

Opposition to incineration is especially strong in Ringaskiddy in Cork Harbour. Here, a toxic incinerator is planned to deal with all of the country's toxic waste. Residents in this area are strongly opposed to the development on the grounds that their local environment and health will be badly damaged by such a venture.

Large incinerators, such as this one in London, could become a key solution for Ireland's waste problem. Would you support such a solution?

Recycling: the sustainable solution to waste?

The volume of waste and its pollution of the environment in Ireland is not sustainable. A *National Waste Management Strategy*, therefore, calls for local authorities and communities to become more committed to recycling/reusing waste materials.

Currently, however, Ireland is one of the least committed to recycling within the EU (Figure 22.1). This has to change, and the Government aims to recycle some 45 per cent of waste by 2010.

Galway City Council uses different coloured bins to collect sorted, clean and renewable waste. The amount of waste sent to landfill fell by 50 per cent between 2001–2003.

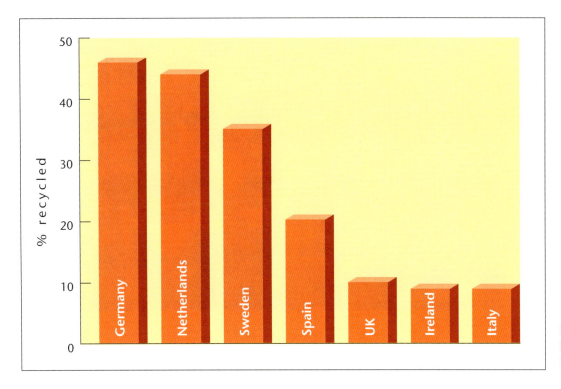

Fig. 22.1 Recycling of household waste in selected EU countries

Class activity

Study Figure 22.1.
1. What percentage of Ireland's waste is recycled?
2. What countries show the highest percentage? Does this surprise you?
3. What is Ireland's target percentage for recycling in 2010?
4. Suggest ways in which this target can be met. Do you think this target is reasonable?

RADIOACTIVE POLLUTION: AN INTERNATIONAL PROBLEM

The nuclear power industry uses small amounts of **uranium** to generate electricity. In the process, radioactive particles are created. These are dangerous to all living organisms. Radioactive particles also remain contaminated for up to 30,000 years. The threat of radioactive pollution to the world's environment will not be resolved quickly.

There are *three* sources of concern over radioactive pollution:

1. **Discharge**, or the escape of radioactive particles from nuclear plants which pollute the water, air and soil.
2. **Disposal** of radioactive water created by nuclear power plants.
3. **Decommissioning** nuclear plants after their normal life span of about 30 years. Isolating and guarding large sites which contain contaminated buildings and material will be difficult, expensive and a long-term commitment.

International concerns over nuclear pollution relate mainly to the *discharge* and *disposal* of radioactive particles from nuclear plants. These are linked to:

● accidents
● 'normal' leakages.

Large demonstration over nuclear waste from the Bradwell nuclear power plant in Britain. Why are communities so concerned over nuclear waste?

Nuclear Accidents: the Chernobyl experience

On 26 April 1986, the world's worst nuclear accident occurred at the Chernobyl nuclear plant near Kiev in the Ukraine (Figure 22.2). The roof of one of its reactors was blown off in an explosion. As a result, some 190 tonnes of radioactive material were released into the atmosphere.

Radioactive material continued to be released over a 10-day period. While most fallout occurred close to Chernobyl, lighter materials were transported long distances by the wind. As a result of changing wind directions and rainfall patterns, the **radioactive fallout was spread over a large area of Europe** (Figure 22.2).

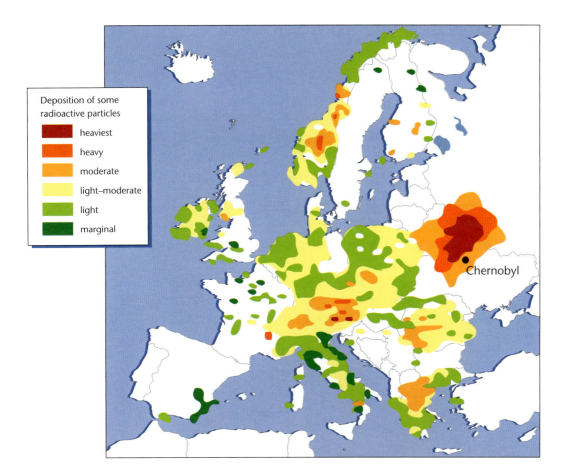

Figure 22.2 Distribution of some radioactive material from the Chernobyl accident

Deposition of some radioactive particles

- heaviest
- heavy
- moderate
- light–moderate
- light
- marginal

Chernobyl

Class activity

Study Figure 22.2.
1. In what country is Chernobyl?
2. Where does the heaviest radioactive pollution occur? Why?
3. Describe the overall pattern of radioactive pollution.
4. Does this pattern support the view that 'pollution knows no boundaries'?

The Chernobyl nuclear power plant following the explosion, which the United Nations described as the worst environmental disaster in human history. Why do you think it was described in this way?

The local effects of the Chernobyl nuclear accident

- The area around Chernobyl is the most radioactive environment in the world.
- Two thousand towns and villages remain uninhabited.
- Large areas of farmland and forest remain contaminated. Food produced from such areas have higher-than-normal levels of radiation and are a health risk.
- Seven million people have been affected by radioactive pollution.
- Cancer cases are abnormally high, especially in children.

143

Radioactive Leakage from Sellafield

The Sellafield nuclear power plant is located on the north-west coast of Britain (Figure 22.3). This extensive operation is concerned with the **storage** and **reprocessing** of nuclear waste.

Sellafield reprocesses both British nuclear waste and used nuclear fuel which is imported from countries such as Germany and Japan. The end products are plutonium and enriched uranium, which can be used as a fuel for nuclear power plants.

This reprocessing plant has a long history of '**minor accidents**' which have resulted in the **leakage of radioactive materials** into the environment. Much of this involves liquid waste which finds its way into the Irish Sea. Largely as a result of this, the **Irish Sea** is regarded as the **most radioactive sea in the world**.

> Importing highly toxic nuclear waste by ship through the Irish Sea is a cause of great concern. A shipping accident would result in massive pollution of the Irish Sea.

The Sellafield nuclear reprocessing plant in north-west England. How does this plant pollute the Irish Sea? Is this of concern to Ireland?

Fig. 22.3 Location of the Sellafield nuclear plant

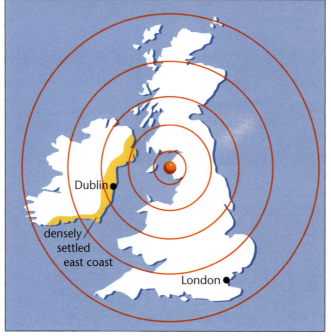

Dublin

densely settled east coast

London

Class activity

1. Is Dublin located nearer to Sellafield than London?
2. What influence do you think this distance factor has on decision-making regarding Sellafield?
3. How could a nuclear accident at Sellafield affect Ireland's densely settled east coast?

POLLUTION AT THE GLOBAL SCALE: GLOBAL WARMING

Since the industrial revolution, average world temperatures have increased. This is referred to as **global warming**. Its main cause is the pollution of the atmosphere by **greenhouse gases** due to the large-scale burning of fossil fuels. This produces what is known as the **greenhouse effect** (Figure 22.4).

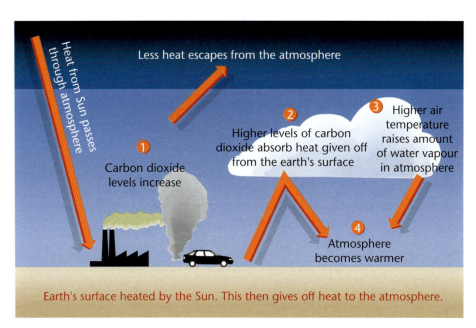

Heat from Sun passes through atmosphere

Less heat escapes from the atmosphere

② Higher levels of carbon dioxide absorb heat given off from the earth's surface

③ Higher air temperature raises amount of water vapour in atmosphere

① Carbon dioxide levels increase

④ Atmosphere becomes warmer

Earth's surface heated by the Sun. This then gives off heat to the atmosphere.

Fig. 22.4 The greenhouse effect

The main greenhouse gas is carbon dioxide (CO_2).

Class activity
Study Figure 22.4.
1. What are the main sources of CO_2 emissions?
2. Why do greenhouse gases cause temperatures to rise in the earth's atmosphere?

TRENDS IN CO₂ EMISSIONS

Emissions of CO_2 have increased strongly since the 1950s (Figure 22.5). These increases have been due to:

● continued and large-scale growth within the developed world economies
● the spread of industrial and urban development into less-developed world regions
● high levels of personal use of energy in developed societies (Figure 22.6)
● rapid growth of population in developing countries. (Why should this be a major threat to global warming? Hint: see Figure 22.5)

The USA, with only 5 per cent of the world population, discharges 25 per cent of global CO_2 emissions.

Burning of wood for heating and cooking in less-developed countries adds to CO_2 emissions. At present the amounts are relatively low, but this will increase rapidly if these people follow the example of the developed world. Why?

The large-scale use of fossil fuels in core countries of Europe and North America is a major cause of global warming.

145

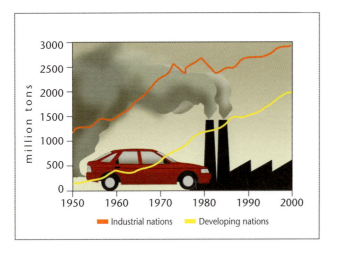

Fig. 22.5 Total carbon dioxide emissions by the industrialised and developing countries, 1950–2000

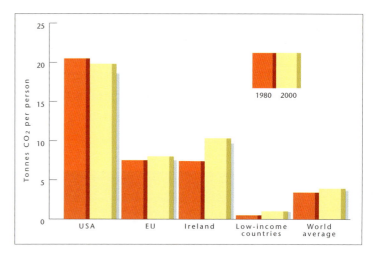

Fig. 22.6 Average amount of CO_2 emitted per person

Class activity

Study Figures 22.5 and 22.6.

1. Describe and account for the trends in CO_2 emissions since 1950.
2. Estimate the CO_2 emissions per person in the USA, world average and low-income countries.
3. Why are emissions per person so high in high-income countries?
4. Why are emissions per person so low in low-income countries?
5. Why are emissions per person in Ireland a cause for concern?

SOME CONSEQUENCES OF GLOBAL WARMING

If levels of greenhouse gases continue to increase, the earth's temperature could rise by 3°C· in the twenty-first century. This could have several important consequences:

- **Sea levels will rise** as polar ice caps melt quickly.
- **Climate changes** will occur and will impact on agricultural land use. For example, droughts, and therefore famines, could become more common in sub-Saharan Africa.
- **Human health** problems could increase in temperate latitudes as tropical diseases, such as malaria, might spread into these regions due to higher temperatures. More intensive sunshine could also cause an increase in cases of skin cancer.

Why would a rise in sea level be disastrous for low-lying coastal regions such as the Ganges Delta? Which country in western Europe would be most affected?

A Solution to Global Warming: the Kyoto Protocol

In 1997, a major international conference on the threat of global warming was held at Kyoto in Japan. Arising from this, an agreement called the **Kyoto Protocol set a target of reducing the 1990 global level of greenhouse gases by 5 per cent by 2012**. This was to be achieved by:

- reducing the burning of fossil fuels
- promoting the use of clean, alternative fuels
- reducing the process of deforestation occurring in less-developed world regions (see pages 162–6).

The USA has refused to sign the Kyoto Protocol. Why does this make it almost impossible to meet the Kyoto target (see Figure 22.6).

Ireland and the Kyoto Protocol

Growth of the Celtic Tiger economy has helped make Ireland one of the worst polluters of greenhouse gases in the EU (Figure 22.6). In 2002, however, Ireland signed the Kyoto Protocol. The country is now committed to **reduce its 1990 levels of CO_2 emissions by 13 per cent by 2012**. This will involve:

- reducing the burning of fossil fuels, especially coal at the Moneypoint power station on the Shannon Estuary
- increasing the use of alternative energy sources, especially wind power (see pages 137–8)
- introducing a carbon tax to make the use of fossil fuels more expensive
- increasing the area of forestry.

TEST YOURSELF AT
my-etest.com

CHAPTER 23
SUSTAINABLE ECONOMIC DEVELOPMENT AND THE ENVIRONMENT

KEY IDEA!

Sustainable economic development takes a long-term view of development and reduces its negative impact on the earth's natural resources.

See Chapter 25 for the problems of deforestation and desertification.

Sustainable economic development means meeting the needs of the present without limiting the ability of future generations to meet their own needs.

Economic development has brought many benefits for humanity. It has also brought problems for past, present and future generations. Our over-use of non-renewable resources, such as fossil fuels, has polluted the environment (e.g. acid rain, global warming), while also reducing the reserves available for future generations. In addition, misuse of renewable resources limits their potential for long-term development.

An alternative model of development is needed. It should take a long-term view of development and reduce the negative impact of economic growth on the environment. This is the model of **sustainable economic development**.

This chapter looks first at the growing significance of environmental planning in Ireland. It then uses *two* examples of how the past and present use of natural resources can affect future development:

- fish stocks in Ireland
- mining in Ireland.

ASSESSING THE ENVIRONMENTAL IMPACT OF ECONOMIC DEVELOPMENT: THE IRISH EXPERIENCE

The United Nations' **Earth Summit** held at Rio de Janeiro in 1992 focused attention on the need for the sustainable development of the earth's economy and environment. One difficulty linked to this approach is our lack of knowledge of the global environment and the potential impacts that development projects can have on the environment. To overcome this problem, **Environment Impact Studies** (EIS) have been introduced.

Environmental Impact Studies (EIS) in Ireland

These now form a vital part of national and county development plans in Ireland. So, before any major development project can proceed, an EIS has to be undertaken. This:

- is made by independent researchers
- assesses and reports on the state of the environment
- evaluates the costs and benefits of the project
- assesses its likely impacts on the environment.

All major infrastructure projects, such as motorways, must undergo an EIS. Explain some important effects that the construction of a motorway may have on the local environment.

The EIS, therefore, evaluates the likely environmental impacts of a project *before* the project is allowed to start. As a result, modifications or restrictions can be imposed on developers in order to reduce any future negative impacts on the environment.

The Role of the Environmental Protection Agency (EPA)

To further promote sustainable economic development in Ireland, the **Environmental Protection Agency** (EPA) was established in 1993.

The EPA closely monitors the state of Ireland's environment, and advises the Government on matters of environmental policy.

One of the main purposes of the EPA is to *licence and control all large-scale activities which have the potential to cause significant environmental pollution*. So, for example, any industry or waste-disposal site wanting to operate in Ireland must apply for an *integrated pollution control* licence. This details the levels of pollution within which they can operate. Industries must report their ongoing emission levels to the EPA which monitors them to ensure they stay within their limits. In this way, economic development should proceed with only limited impacts on the environment.

The mission statement of the EPA is to promote and implement the highest practical standards of environmental protection and management for sustainable and balanced development.

Maintaining a clean environment in Ireland is vital for some 155,000 jobs in farming, forestry, fishing and tourism.

Cork Harbour is the chemical capital of Ireland. All these chemical plants must have an EPA licence to operate. Why is this important?

149

Sustainable Development and Ireland's Fish Stocks

Ireland's territorial waters (see Figure 17.1) include some of the richest fishing grounds in the European Union. For a long time, this natural resource remained underdeveloped. Since the 1970s, however, pressures on fish stocks have increased rapidly. The result has been *overfishing* and the *depletion* of fish stocks. This is *not* sustainable economic development.

The Irish-owned *Atlantic Dawn* is one of the world's largest fishing and fish-processing vessels. Due to its size and modern equipment, it can stay at sea for long periods and also fish the oceans well outside EU waters. Do you think the future of Ireland's fishing industry lies in such vessels?

The great growth in fishing activities and pressures on fish stocks in Irish waters are due to:

Refer to Chapter 18 in *Our Dynamic World I* to review developments in Irish fishing.

- an **increase in the size and efficiency of fishing fleets** that are able to catch larger volumes of fish. In Ireland, for example, fish landings increased almost four-fold to some 300,000 tonnes from 1973 to 2001.
- **unrestricted access to Irish waters for the large fishing fleets from other EU member states**. This is the *main cause* of overfishing, since foreign trawlers account for some 80 per cent of the value of all fish caught in Irish waters.

As a natural resource, fish can sustain a reasonable rate of exploitation *providing there is no prolonged overfishing*. Fish stocks can be sustained in two ways:

1. Allow the spawning stock to reproduce at an effective level.
2. Allow young fish to mature to add to the breeding population.

Remember the impact of the CFP on Irish fishing in Chapter 17?

The Common Fisheries Policy (CFP) of the EU has tried to encourage this in *three* ways:

- establishing a **total allowable catch (TAC)** and a national quota for each species of fish caught in EU waters
- issuing **conservative measures** to protect young and spawning fish, such as preventing fishing in certain areas called *fishery exclusion zones*. Several of these occur in Irish waters.
- **controlling and enforcing** the TACs and conservation measures.

These measures have *not* been successful. National governments and fishing organisations have argued successfully to increase fish quotas and TACs. These have, therefore, been set at too high a level for fish stocks to replace themselves (Figure 23.1). In addition, control and enforcement of conservation measures have been difficult. *The result has been overfishing and depletion of fish stocks in EU and Irish waters.*

Ireland's territorial waters are rich fishing grounds, but overfishing has led to the large-scale depletion of fish stocks. Explain why.

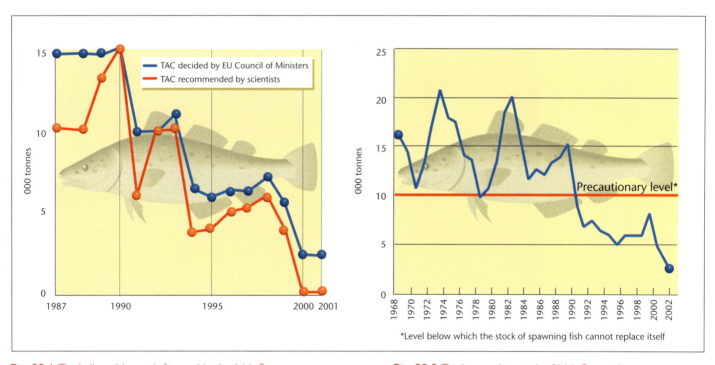

Fig. 23.1 Total allowable catch for cod in the Irish Sea

Fig. 23.2 Total spawning stock of Irish Sea cod

Class activity

Study Figures 23.1 and 23.2.
1. Describe the trend for the TAC in cod. Why did this occur?
2. How do the TAC levels compare with scientific recommendations? What does this mean for cod stocks?
3. In what year did Irish Sea cod fall below its precautionary level?
4. What has been the main trend for spawning cod since the early 1980s?
5. Can you see any linkage between the two graphs?

By 2000, two-thirds of fish stocks in EU waters were approaching commercial extinction.

151

Problems of overfishing by Irish and foreign trawlers in Irish waters began in the 1970s. Herring stocks in the Celtic Sea were almost wiped out and this led to the closure of this area for herring fishing between 1977 and 1982. This caused fishing fleets to put pressure on other species, such as mackerel and cod (Figure 23.2).

Most fish species in Irish waters are now under threat. For an industry which depends on a renewable resource, this is not sustainable. **Unless much stronger and more effective policies are put in place, and are accepted by the fishing industry, one of Ireland's richest renewable resources will be lost.** This will be a disaster for the environment and for coastal fishing communities.

The future of fishing communities, such as Killybegs, depends largely on the sustainable development of fish resources in EU waters. Explain why, especially in terms of employment.

Mining and Environmental Impact

Mining refers to the activity of digging, generally below ground level, to extract non-renewable mineral resources, such as coal and copper. In the past, this activity gave rise to major problems for the environment around the mine site. These include:

● a huge build-up of waste which litters the landscape
● the pollution of rivers and ground water
● large amounts of dust in the atmosphere.

Communities in the south Wales coalfield usually developed around a local coal mine. Waste material from the mines gave rise to polluted environments. Why was regional planning important for such areas?

In addition, mining generally requires large numbers of workers. Communities are, therefore, built up in close proximity to the mine and become dependent on it for their livelihood.

When mineral resources become exhausted, or too costly to extract, the mine closes. This leaves:

- a *despoiled landscape* that repels new investment
- mining communities with *high unemployment*
- governments needing to invest large amounts of money, via *regional planning*, to restore or rehabilitate the environment and encourage new industries to relocate to mining communities.

This is not sustainable economic development.

Since the 1970s, however, the mining industry in developed countries has had to be more aware of its environmental impacts. This means the industry is required to reduce pollution levels and also to restore the quality of the environment if pollution occurs. In this way, mining now has a less negative impact on the local environment and communities.

Case Study: Tara Mines

In 1977, Europe's largest zinc and lead mine began production near *Navan*, Co. Meath. *Tara Mines* extracts the ore from extensive underground workings, and processes it into almost pure zinc and lead called **concentrate**. The concentrate is transported by rail to Dublin Port for export.

By the 1970s, planning restrictions were in place in Ireland to reduce the environmental impact of mining operations. This was particularly the case for Tara Mines since it was sited near urban centres and in a rich agricultural region. It was also close to the Blackwater, one of the country's prime fishing rivers.

Planning conditions

- The mining site was to be screened by trees to reduce visual impact.
- Noise and air pollution closely monitored.
- Large quantities of mining waste, or tailings, were carefully managed.
- Water used in the operation was purified before being released into the Blackwater.

> Remember the problems of the Sambre Meuse Coalfields in Chapter 16 of *Our Dynamic World 1*?

> Initial reserves were estimated at some 70 million tonnes of ore, although these have been revised upwards.

> Some 7 per cent of the costs of development was given to conservation measures.

Site of Tara Mines, the largest zinc and lead mine in Europe, processing over 2.5 million tonnes of ore a year. Use evidence from this photograph to discuss whether the development of Tara Mines has had a large-scale and negative impact on its environment. Then compare the photo of Tara Mines with the photo on page 153. Which suggests more sustainable economic development?

Disposal of waste is the main problem for Tara Mines (Table 23.1). Each year, 1 million tonnes is returned underground to backfill areas that have been mined. Most waste, however, is piped to a tailings pond located 5 km from the mine. Here, the solid particles settle to the bottom of the pond, leaving clear water above.

The waste from processing the ore is called *tailings*. It is a mixture of water and solid particles.

Total production at Tara Mines, 1977–2003	
Ore processed	55.2 million tonnes
Zinc concentrate	7.5 million tonnes
Lead concentrate	1.4 million tonnes

Table 23.1

Class activity
Study Table 23.1.
What do these figures tell you about the waste problem for Tara Mines?

The water from the tailings pond is recycled for use in the mine. Over time, the level of solid tailings builds up. Eventually, the water is drained off the tailings leaving new land. This *rehabilitated land* can then be used for grassland or left as a natural wetlands.

In the ways described above, the surrounding environment is not extensively damaged. The environment is rehabilitated so that it will not repel new developments when the mine closes. This is important for the 650 workers and communities that depend on the mine. As a result, *modern mining practices and concerns with environmental impact can be seen as a form of sustainable economic development.*

Tara Mines has an integrated pollution licence from the EPA.

TEST YOURSELF AT

CHAPTER 24
ECONOMIC DEVELOPMENT OR ENVIRONMENTAL PROTECTION: A CAUSE FOR LOCAL CONFLICT

KEY IDEA!

Conflicts of interest can occur within local communities over the extent to which their natural environments should be exploited for economic gain or protected to maintain its quality.

Economic development is generally viewed positively by most people, especially if their quality of life is improved. It can, however, cause a conflict of interests within local communities. This generally focuses on the priority that should be given to protecting the environment as opposed to promoting its use for immediate economic gain.

This chapter reviews local conflicts of interest that have arisen in Ireland over:

- large-scale development of fish farming
- peat extraction or bogland preservation.

Another example would be the conflict over developing heritage tourism in the Burren, such as the Mullaghmore visitor centre.

FISH FARMING IN IRELAND

Fish farming has developed rapidly in Ireland since the early 1980s (Figure 24.2). This recent growth has been due to:

1. Restrictions placed on total fish catch at sea under the CFP of the EU (see Chapters 17 and pages 150–53).

2. Increasing market demand for fresh fish and fish products.

3. Excellent environmental conditions for fish farming, especially along Ireland's western coastline. These include:

   - pollution-free water due to the absence of large-scale, urban-industrial development
   - many sheltered bays and estuaries
   - regular tidal flows to help keep the water clean.

Fish farming is the cultivation of fish under controlled conditions, for example in large fish cages.

Fish farming on the Beara Peninsula in West Cork. Using this photograph, suggest some of the main advantages and disadvantages of promoting fish farming in such areas.

Economic Interests Promoting Fish Farming

Several economic arguments have been used to justify the large-scale development of fish farming along Ireland's western seaboard.

- Remote coastal communities have **high rates of unemployment and emigration**. In addition, the introduction of fish quotas and conservation measures under the EU fishers policy reduced one of the main natural advantages possessed by small fishing communities.

- These **coastal communities need to be supported**, and both the Irish Government and the EU identified fish farming as having good prospects for development. *Generous incentives* were therefore offered to help set up fish-farming operations. Most are located along the western coastline, and especially in counties Donegal, Mayo, Galway, Kerry and Cork (Figure 24.1).

- The **natural environments** of these areas are very good for fish farming. In addition, the *fishing tradition* and use of local, renewable resources help to *reduce costs* of production.

- Despite government support, the costs of setting up large fish farms are high. Companies are, therefore, encouraged to **rear as many fish as possible** to offset costs and make a good profit.

- The **creation of new jobs and wealth** was the main argument for developing fish farming. By 2002, over 1,800 jobs had been provided in fish farming.

- The success of fish farming also generates **spin-off industries**, and therefore more jobs and wealth for these coastal communities, e.g. building fish cages, preparing fish feed, processing the fish.

The Irish Government is committed to further development of fish farming and to double its output in 2003–2008.

Fig. 24.1 Locations of large-scale fish farming in the west of Ireland. Why are such locations suitable for fish farming?

Areas of large-scale fish farming

Environmental Objections to Fish Farming

Initially, there were few objections to the development of fish farming in Ireland. However, as the *scale* of its development increased, increasing objections have been raised regarding the impact of fish farming on **local environments**.

- **Water quality** declines due to the large-scale use of chemicals to help combat the spread of parasites and disease among the high concentrations of fish. Chemicals are also used to clean the fish cages. These chemicals are dispersed throughout the water bodies in which the fish farms are located.

- **Water pollution** levels also increase as excess fish food, faeces and dead fish fall and build up below the cages. This build-up of material pollutes the water, which was a major initial attraction for fish farming.

- **Diseases** can break out and spread quite easily through the fish in the high-density environments. **Sea lice**, in particular, are a major problem and kill large numbers of fish.

- **Escaping farm-bred fish** can interbreed with native wild fish species. Also, diseased fish and the spread of sea lice, mainly from salmon farms, have caused a **massive decline in native sea trout and in wild salmon**. As a result, there are now far fewer native sea trout and wild salmon found in Ireland's main fishing rivers. (Is this good for angling tourism?)

- The over-development of fish farming can affect the **scenic qualities** of the coastal environments in which they locate. For example, fish cages and pollution washed up on the shoreline can be unsightly.

> Most environmental concerns are linked to the scale of fish farming, and especially salmon, where many thousands of fish are confined in cages within relatively small areas.

> The Federation of Irish Salmon and Sea Trout Anglers has suggested that Ireland's stocks of native sea trout and wild salmon have become victims of ethnic cleansing. What do you think this means?

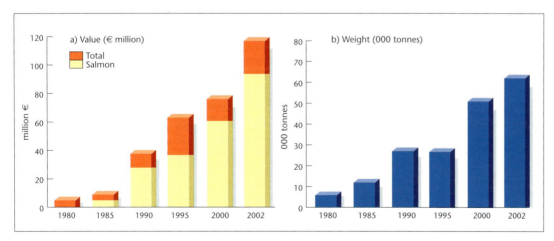

Fig. 24.2 Growth in the output of Irish fish farming, 1980–2002

Class activity

Study Figure 24.2.

1. Estimate the total value and weight of products from fish farming in 2002.
2. How do these values compare to 1980?
3. Explain this growth trend.
4. What is the most important type of fish by value?
5. Do you think this scale of development has been good or bad for coastal communities? Explain your answer.

The Conflict of Interests

The economic interests of fish farmers and the Government are to further develop the industry. This, they argue, will bring more jobs and prosperity to coastal communities.

Many local communities, however, fear that large-scale fish farming has negative impacts on their environment. They say this affects the development prospects of other activities which depend on the local coastal environment, and which could help to diversify their economy. These include:

● tourism

● other forms of fishing, both coastal and inland

● water sports.

Class activity
Discuss ways in which you think fish farming impacts on each of these activities.

Integrated Coastal Zone Management: a solution?

There is an urgent need for effective planning to help resolve the above conflict of interests. This should involve an Integrated Coastal Zone Management scheme. The schemes try to ensure that various interest groups, such as fish farmers, tourists, anglers, are able to use the coastal environment *without* damaging it for others. In this way *both the economic and environmental interests* of local communities are protected, and allows for long-term sustainable development.

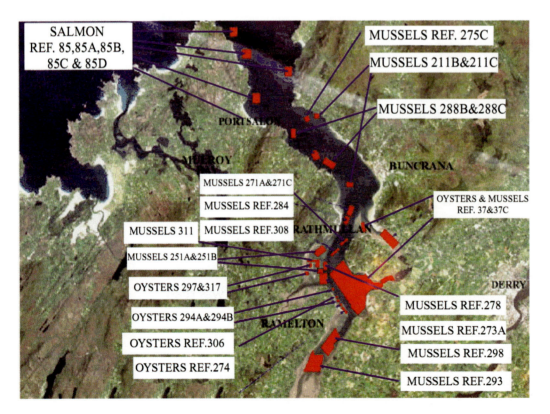

SALMON REF. 85,85A,85B, 85C & 85D

MUSSELS REF. 275C

MUSSELS 211B&211C

MUSSELS 288B&288C

PORTSALON

MILROY

BUNCRANA

MUSSELS 271A&271C

MUSSELS REF.284

OYSTERS & MUSSELS REF. 37&37C

MUSSELS 311   MUSSELS REF.308   RATHMULLAN

MUSSELS 251A&251B

OYSTERS 297&317

DERRY

OYSTERS 294A&294B   RAMELTON

MUSSELS REF.278

OYSTERS REF.306

MUSSELS REF.273A

OYSTERS REF.274

MUSSELS REF.298

MUSSELS REF.293

This map represents fish licences currently active and pending approval on Lough Swilly (2001). Why do you think an organisation called 'Save the Swilly' has emerged to campaign against the scale of fish farming (1,000 acres) in this area?

Fish Kills in Inver Bay, Co. Donegal

Inver Bay is an important location for fish farming and includes three large salmon farms (can you locate it on Figure 24.1?). In July 2003 up to 500,000 farmed salmon were believed to have died in a salmon pen owned by one company. This amounts to some 2,200 tonnes of dead fish. Furthermore, this disaster followed on from the death of 50,000 farmed salmon at the same location in 2002. The fish were allowed to sink to the sea bed and decompose.

The cause of these large fish kills are unclear. What is clear, however, is the threat that such disasters pose for the environment. Furthermore, they emphasise the need for stricter controls over the scale of fish farming occurring in relatively small areas.

IRELAND'S BOGLANDS: AN ISSUE OF HERITAGE?

At one time, Ireland's *boglands* covered large areas of the country (Figure 24.3). They were an important natural resource in that they provided *peat* which could be used as a source of fuel.

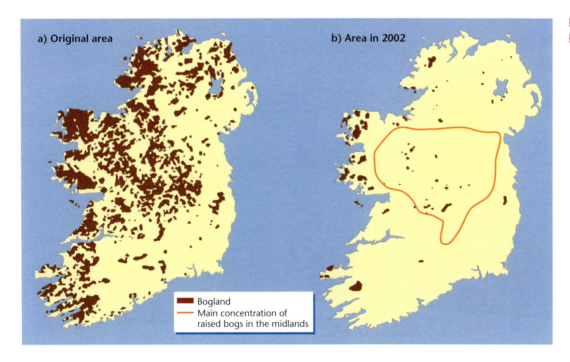

a) Original area

b) Area in 2002

Bogland
Main concentration of raised bogs in the midlands

Fig. 24.3 Changing area of boglands in Ireland

While peat extraction from Ireland's boglands has occurred over many centuries, the rate of removal increased markedly following the 1950s. **Exploitation of peat was seen as important for the country's economic development**. Before the development of its natural gas resources, Ireland had no alternative energy supply, apart from some hep. Dependency on imported energy sources was, therefore, extremely high. This was an economic disadvantage, highlighted by the oil crises of the 1970s, and encouraged the large-scale development of boglands for peat.

The state established Bord na Mona in 1946 to develop the economic potential of peat.

Machine-cutting of peat from raised bogs, such as the Bog of Allen in the midlands, removes large amounts of this natural resource for burning in power stations. Do you think this is the best use for this declining natural resource?

Most peat was extracted by Bord na Mona from the large **raised bogs** of the midlands. Here, the more level terrain and deep layers of peat allowed for strip mining of peat by large machines. Most of the peat was used as a fuel source for peat-fired power stations. This *helped the national economy and also provided many jobs* for communities that became dependent on the peat industry.

On the large areas of **blanket bog** in the west of Ireland, peat was also harvested as an economic resource. Farmers cut the peat as a source of domestic fuel and sold the surplus as a cash crop. It was, therefore, an essential resource for many marginal communities.

As a result of the economic exploitation of peat, only 20 per cent of Ireland's original boglands remain (see Figure 24.3).

This loss of boglands is a source of concern for many people interested in preserving the country's natural environment. Ireland's boglands are regarded as one of Europe's last wilderness areas, possessing many unique qualities. As a result, **they need to be protected as part of our heritage** and not lost for the production of peat:

- Boglands support a great range of rare plants and animals.
- They are important as bird sanctuaries.
- Many archaeological sites and cultural relics are preserved in the boglands.
- They are attractive natural environments which can be promoted to attract tourists.

Remember the formation of peat and the difference between raised and blanket bogs from your Junior Cert?

Preservation of boglands will create some new jobs/wealth through eco-tourism, e.g. bird watching, rambling. Will this help communities adjust to job losses in the peat industry?

Traditional hand-cutting and drying of peat on blanket bogs in the west of Ireland remains an important part of the land use for many upland farmers. Why? Does the photograph show an alternative and renewable land use for such upland areas?

Blanket bog landscape in the Inishowen Peninsula, Co. Donegal. Do you think such landscapes should be protected? Do you think the farmer living in the cottage would have the same opinion?

A **conflict of interest** has emerged over the conservation or exploitation of boglands. The interests of communities that are dependent on peat extraction have to be protected, but equally the loss of Ireland's boglands will leave the country's natural environment in a poorer state.

Farmers enraged as EU orders ban on turf-cutting

More than 6,000 farmers have been ordered to stop cutting turf on their land. The move follows criticism from European Union partners over Ireland's failure to enforce laws protecting about 200,000 acres of bog as special areas of conservation.

Farmers' representatives say the crackdown is the latest attack on a way of life already under siege.

Because of their unique chemical properties they are home to rare flora and fauna, which environmentalists are anxious to protect. They are among the last remaining wildernesses in Europe and have been proven to serve a historical purpose by preserving tools and even human remains for thousands of years.

For the farmers, however, the wetlands are a source of fuel. Turvey rights, allowing families to harvest turf from local bogs for use during winter, have been passed down from generation to generation.

Turf provides cheap fuel and additional income to farmers around Galway, Roscommon, Mayo, Kerry, Donegal and large parts of the midlands. Those who do not use their rights often sell them on to professional turf cutters.

The farmers' association says it can understand why the Government is trying to clamp down on industrial harvesting by Bord na Mona and the ESB. However, a spokesperson added: 'I cannot see in what way the harvesting of small amounts of turf on a hillside bog can be especially harmful to the environment. It has been going on for hundreds of years and the wildlife has prospered.'

Sunday Times, 15 September 2002

Class activity

Read the newspaper extract.

1. How many acres of bogland are to be conserved?
2. Why are the bogs to be conserved?
3. Why do farmers object to the loss of turf-cutting?
4. Do you support the farmer or the EU over the issues of cutting turf? Why?
5. Do you have the same attitude as Bord na Mona? Why?

TEST YOURSELF AT
my-etest.com

CHAPTER 25
ECONOMIC DEVELOPMENT OR ENVIRONMENTAL PROTECTION: GLOBAL CONCERNS

 **KEY IDEA!**

At the global level, large-scale exploitation of sensitive environments create major problems, such as deforestation and desertification.

The previous chapter introduced conflict of interests at a local level between people favouring economic exploitation of the environment as opposed to those favouring its conservation. This chapter reviews similar conflicts, but these are on a much larger scale and give rise to major global concerns. *Two* examples of global conflict are highlighted:

- deforestation
- desertification.

DEFORESTATION

The cutting of trees has long been a characteristic of development, as people create more land for farming and settlement, and wood for fuel.

For example, large areas of Europe were cleared of trees in the past, as in Ireland where forests now cover only 8 per cent of the landscape. Developed countries, however, now recognise significant economic and environmental benefits of forests. As a result, *reforestation* rather than deforestation is more typical in the developed world (Figs. 25.1 and 25.2).

Deforestation means the large-scale cutting and clearing of forests for wood products and to create new land for agriculture and other land uses.

Can you think of any economic and environmental advantages of increasing forest areas?

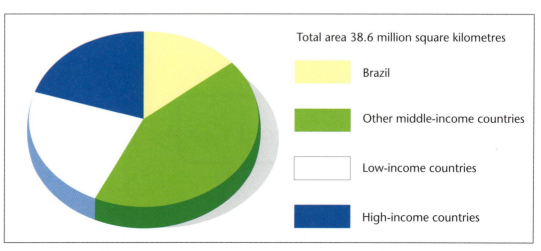

Total area 38.6 million square kilometres

- Brazil
- Other middle-income countries
- Low-income countries
- High-income countries

Fig. 25.1 The world's forested area

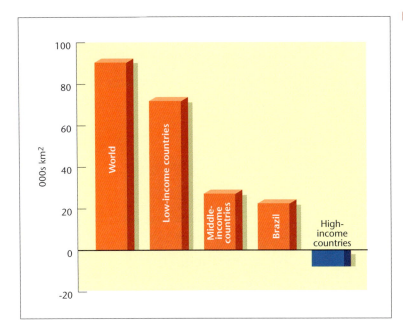

Fig. 25.2 Average annual deforestation, 1900–2000

Class activity

Study Figures 25.1 and 25.2.

1. Estimate the share of world forest that is located in high-income countries. Why is it so low?
2. Where are most of the world's forests located?
3. Estimate yearly rates of global deforestation in the 1990s.
4. Which global region shows greatest rates of deforestation? Why?
5. What is unusual about trends in high-income countries? Explain.

Deforestation continues to occur on a vast scale in the less-developed world, and *especially in tropical regions*. Exploitation of these valuable natural resources is seen by many people as offering important economic benefits. Opposed to this view, however, are a growing number of people who see *large-scale deforestation of the tropics as not only a natural, but also a global, environmental disaster*.

About 2 billion people living in developing countries still rely on wood and dung for fuel. This contributes significantly to deforestation and desertification.

Economic Interests Favouring Deforestation

Economic interests are often used to justify large-scale deforestation in less-developed countries, such as Brazil (Figure 25.3). These include the following:

● **development policies of national governments**. Deforestation provides wood products and increases food output as land becomes available for farming. These products are usually exported to earn income for the country.

● **national and international companies,** which see major profit opportunities from deforestation, such as logging companies and food producers. In *Brazil*, for example, food producers are often involved in large ranches that are established on new grasslands created in deforested areas. Beef cattle are raised to export low-cost meat to the developed world markets.

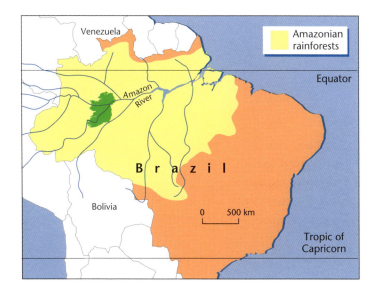

Fig. 25.3 Tropical rainforests cover 5.3 million km², almost two-thirds of Brazil is covered by tropical rainforests. Each year, however, 22,000 km² are deforested, an area almost one-third the size of Ireland. (Note the size of Ireland drawn on the map of Brazil.)

163

- the **rapidly growing population and large numbers of landless peasants,** who put increasing pressure on available land in many less developed countries. Clearing forested areas is often seen as an easy option to meet the 'land hunger' of the population.
- **consumers in the developed world,** who provide a major market for tropical wood products and cheap food. For countries with a *debt crisis,* increasing export earnings is vital.

Large-scale deforestation in Brazil. Is this practice in the long-term interests of Brazil and the global environment?

Economic and Environmental Interests Opposing Deforestation

A growing body of both local and global opinion opposes large-scale deforestation for economic and environmental reasons.

At the **local level**, deforestation in tropical areas is linked to the longer-term loss of productive land as well as natural environment:

- The removal of tree cover exposes soil to heavy tropical rains. These *wash away more fertile upper soil layers.* Nutrients are also leached out of the soil.
- The absence of large quantities of tree litter *removes the means of maintaining soil fertility,* e.g. decomposing leaves.

Large areas of tropical rainforest are burned to create new farmland for local farmers and also to create extensive cattle ranches. Does this create fertile land for long-term development?

Soil and land are considered to be a renewable resource. Is this the case under deforestation?

164

- Initially, after burning and clearing the forests, soil fertility is quite high. The above processes and *demands of farming, however, quickly exhaust the soil.*
- Farmers (and loggers) then have to clear more forest. *This is not a sustainable form of economic development.*

The scale of deforestation is also **an issue of increasing global concern**. This relates principally to *loss of habitat and global warming*:

- Tropical rainforests support a vast range of plant and animal species. Loss of this habitat causes *a decline in the biodiversity of the earth*. This is not only an environmental issue, but it also has importance for human health and development. Many *pharmaceutical drugs* have originated out of research on species found only in these rainforests. Great potential exists for more discoveries, but only *if we protect these vital global habitats.*
- Loss of habitat and exposure to modern cultures *threaten the survival of primitive tribes* who live in these rainforests. Their *human rights* to live in their home environment need protection from powerful economic forces.
- Perhaps the most important global issue centres on the contribution of deforestation to *global warming*. Tropical rainforests are considered to be the '*lungs of the world*'. This is because these vast forests absorb large quantities of CO_2 from the atmosphere and convert it into oxygen.
- Deforestation reduces this natural process. Furthermore, the burning and decomposition of trees adds to the amount of CO_2 in the atmosphere. *Conserving the rainforests, therefore, has a vital role to play in protecting the global environment.*

Tropical rainforests, as in Brazil, cover only 6 per cent of the earth's surface, but contain over 50 per cent of its plant species. Does this give both an economic and an environmental importance to these forests?

Remember global warming? See pages 145–7.

Possible Solutions for Sustainable Development

The short-term economic exploitation of tropical rainforests has to be replaced by approaches that encourage long-term, sustainable development. These could include:

1. Promoting the renewable resources of the rainforests as a source of wealth and employment, e.g. harvesting food and medicinal products, selective cutting of trees and replanting.
2. Establishing national parks in the rainforests to attract high-income tourists to visit these exotic environments.
3. Rewarding developing countries that conserve large areas of rainforests in the interests of protecting the global environment. For example, the *Debt for Nature Scheme* allows developing countries to offset some of their debt by protecting large areas of tropical rainforest.

Do you think this Debt for Nature Scheme is a good idea to help development and protect the environment?

165

An area in the Amazon Basin. The soil has been exhausted after only five years of extensive livestock grazing and now supports little vegetation.

Traditional tribes in the Amazon rainforest have lived in harmony with nature for centuries through hunting and gathering food.

Class activity

Study the photographs above.

1. Suggest reasons why deforested tropical areas often lead to poverty for people attracted to such areas.
2. Does the way of life of traditional tribal peoples in tropical rainforests suggest a better way of using these vast renewable resources? Explain.

DESERTIFICATION

Desertification is a process involving the spread of desert conditions, usually into adjacent semi-arid environments. This causes the loss of fertile soil and vegetation cover, and therefore reduces the ability of extensive areas to support life. Furthermore, once the process of desertification begins, it is extremely difficult (and expensive) to stop.

This is a large-scale and growing global problem. The global region *worst affected by desertification*, however, lies along the southern edge of the Sahara Desert. This is called the *Sahel* (Figure 25.4).

Some estimates suggest that in the Sudan, for example, the desert frontier has been advancing at 5.5–9km a year.

Causes of Desertification: the Sahel

Desertification is caused by both natural and human processes.

Natural factors

Rainfall totals are very low and unreliable in the Sahel. It also experiences long periods of drought.

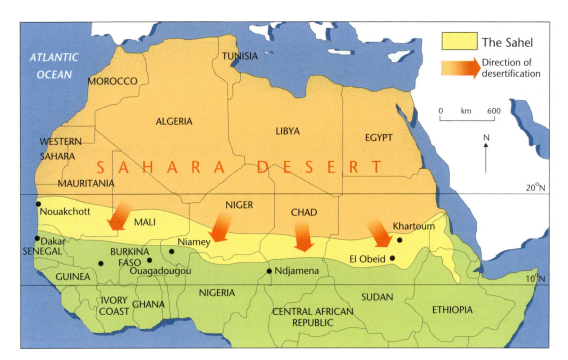

Fig. 25.4 The Sahel

Human factors

These centre on the *over-use* by a growing population of the fragile environment in this frontier zone between desert and semi-desert. This involves:

- overgrazing
- over-cultivation
- deforestation.

These processes have occurred as a result of conflict between local and more global interests over the use of this region's land resource. In effect, **national and global economic interests have over-ridden local and more traditional land uses** (Table 25.1). The result is desertification.

> Deforestation occurs as more land is needed for farming and wood for fuel.

Table 25.1 Economic versus environmental interests in the Sahel

Global and national economic interests	Local and traditional interests
• Global demand for primary products, such as groundnuts (peanuts), cotton, encourages intensive farming.	• Concentration on a variety of food crops grown on a rotation basis, and including a fallow period, prevents soil exhaustion. No over-cultivation.
• National governments promote cash crops for export to help finance development and repay global debt.	• Nomadic herders graze a variety of animals over extensive areas to avoid overgrazing.
• Falling world prices demand a further increase in output. This pushes intensive farming into more marginal environments.	• Traditional land uses are, therefore, well adapted to the difficult environment.
• Soils cannot support intensive farming. Fertility declines, causing farming to move into more marginal land.	• Traditional farmers are pushed into more marginal areas by economic interests. As their populations increase, they abandon traditional methods and begin to over-cultivate/overgraze.

In Niger, intensive farming occurs along the Niger River to produce crops mainly for export. (Note: few farm buildings.) This forces people to move to less-fertile areas where they grow food crops such as millet. Why does this lead to desertification?

Nomadic herders in the Sahel zone in Niger. Note the number of animals and the ground cover. Does this help explain desertification?

Consequences of Desertification in the Sahel

1. Over-use of soil and removal of vegetation cover expose the dry and exhausted soil to wind erosion. Winds carry away the soil as *dust storms, leaving a largely barren landscape* to become part of the advancing Sahara Desert.

2. Failure of the land to provide adequate harvests and grazing, especially in years of drought, results in large-scale famine. Millions die of hunger or disease as the health of the population declines.

3. As the Sahel fails to support its population, millions are *forced to migrate southwards.* This can lead to overpopulation in reception regions. One result is an over-use of land to feed these people which encourages the *further spread of desertification*.

4. Large amounts of financial aid and improved education are vital to check desertification, e.g. to plant trees/shrubs to prevent soil erosion, and to encourage more traditional and sustainable land use.

Peasants planting trees in order to prevent soil erosion. Why would the trees help check desertification?

At the global level, one estimate suggests that desertification threatens 35 per cent of the earth's land and 20 per cent of its population. The future well-being of the earth and its population cannot afford to lose such a vast renewable resource if sustainable economic development is to occur.

TEST YOURSELF AT
my-etest.com